Mat... ud

Acknowledgements

Jane Prothero and Woodlands Primary School, Leeds

Karen Holman and Paddox Primary School, Rugby

Heather Nixon and Gayhurst Primary School, Buckinghamshire

John Ellard and Kingsley Primary School, Northampton

Jackie Smith, Catherine Torr and Roberttown CE J & I School, Kirklees

Wendy Price and St Martin's CE Primary School, Wolverhampton

Helen Elis Jones, University of Wales, Bangor

Ruth Trundley, Devon Curriculum Services, Exeter

Trudy Lines and Bibury CE Primary School, Gloucestershire

Elaine Folen and St Paul's Infant School, Surrey

Jane Airey and Frith Manor Primary School, Barnet

Beverley Godfrey, South Wales Home Educators' Network

Kay Brunsdon and Gwyrosydd Infant School, Swansea

Keith Cadman, Wolverhampton Advisory Services

Helen Andrews and Blue Coat School, Birmingham

Oakridge Parochial School, Gloucestershire

The Islington BEAM Development Group

Published by BEAM Education

Maze Workshops

72a Southgate Road

London N1 3JT

Telephone 020 7684 3323

Fax 020 7684 3334

Email info@beam.co.uk

www.beam.co.uk

© Beam Education 2006

ISBN 1 903142 88 1

British Library Cataloguing-in-Publication Data

Data available

Edited by Marion Dill

Designed by Malena Wilson-Max

Photographs by Len Cross

Thanks to Rotherfield Primary School

Printed in England by Cromwell Press Ltd

Contents

Introduction

Language plays an important part in the learning of mathematics – especially oral language. Children's relationship to the subject, their grasp of it and sense of ownership all depend on discussion and interaction – as do the social relationships that provide the context for learning. A classroom where children talk about mathematics is one that will help build their confidence and transform their whole attitude to learning.

Why is speaking and listening important in maths?

- Talking is creative. In expressing thoughts and discussing ideas, children actually shape these ideas, make connections and hone their definitions of what words mean.
- You cannot teach what a word means – you can only introduce it, explain it, then let children try it out, misuse it, see when it works and how it fits with what they already know and, eventually, make it their own.
- Speaking and listening to other children involves and motivates children – they are more likely to learn and remember than when engaging silently with a textbook or worksheet.
- As you listen to children, you identify children's misconceptions and realise which connections (between bits of maths) they have not yet made.

How does this book help me include 'speaking and listening' in maths?

- The lessons are structured to use and develop oral language skills in mathematics. Each lesson uses one or more classroom techniques that foster the use of spoken language and listening skills.
- The grid on p17 shows those speaking and listening objectives that are suitable for developing through the medium of mathematics. Each lesson addresses one of these objectives.
- The lessons draw on a bank of classroom techniques which are described on p8. These techniques are designed to promote children's use of speaking and listening in a variety of ways.

How does 'using and applying mathematics' fit in with these lessons?

- Many of the mathematical activities in this book involve problem solving, communication and reasoning, all key areas of 'using and applying mathematics' (U&A). Where this aspect of a lesson is particularly significant, this is acknowledged and expanded on in one of the 'asides' to the main lesson.

What about children with particular needs?

- For children who have impaired hearing, communication is particularly important, as it is all too easy for them to become isolated from their peers. Speaking and listening activities, even if adapted, simplified or supported by an assistant, help such children be a part of their learning community and to participate in the curriculum on offer.

- Children who speak English as an additional language benefit from speaking and listening activities, especially where these are accompanied by diagrams, drawings or the manipulation of numbers or shapes, which help give meaning to the language. Check that they understand the key words needed for the topic being discussed and, where possible, model the activity, paying particular attention to the use of these key words. Remember to build in time for thinking and reflecting on oral work.
- Differences in children's backgrounds affect the way they speak to their peers and adults. The lessons in this book can help children acquire a rich repertoire of ways to interact and work with others. Children who are less confident with written forms can develop confidence through speaking and listening.
- Gender can be an issue in acquiring and using speaking and listening skills. Girls may be collaborative and tentative, while boys sometimes can be more assertive about expressing their ideas. Address such differences by planning different groups, partners, classroom seating and activities. These lessons build on children's strengths and challenge them in areas where they are less strong.

What are the 'personal skills' learning objectives?

- There is a range of personal and social skills that children need to develop across the curriculum and throughout their school career. These include enquiry skills, creative thinking skills and ways of working with others. Some are particularly relevant to the maths classroom, and these are listed on the grid on p18.

What about assessment?

- Each lesson concludes with a section called 'Assessment for learning', which offers suggestions for what to look out for during the lesson and questions to ask in order to assess children's learning of all three learning objectives. There is also help on what may lie behind children's failure to meet these objectives and suggestions for teaching that might rectify the situation.
- Each section of four lessons includes a sheet of self-assessment statements to be printed from the accompanying CD-ROM and to be filled in at the end of each lesson or when all four are completed. Display the sheet and also give children their own copies. Then go through the statements, discussing and interpreting them as necessary. Ask children to complete their self-assessments with a partner they frequently work with. They should each fill in their own sheet, then look at it with their partner who adds their own viewpoint.

How can I make the best use of these lessons?

- Aim to develop a supportive classroom climate, where all ideas are accepted and considered, even if they may seem strange or incorrect. You will need to model this yourself in order for children to see what acceptance and open-mindedness look like.
- Create an ethos of challenge, where children are required to think about puzzles and questions.
- Slow down. Don't expect answers straight away when you ask questions. Build in thinking time where you do not communicate with the children, so that they have to reflect on their answers before making them. Expect quality rather than quantity.
- Model the language of discussion. Children who may be used to maths being either 'correct' or 'incorrect' need to learn by example what debate means. Choose a debating partner from the class, or work with another adult, and demonstrate uncertainty, challenge, exploration, questioning ...
- Tell children what they will be learning in the lesson. Each lesson concludes with an 'Assessment for learning' section offering suggestions for what to look out for to assess children's learning of all three learning objectives. Share these with the children at the start of the lesson to involve them in their own learning process.

How should I get the best out of different groupings?

- Get children used to working in a range of different groupings: pairs, threes or fours or as a whole class.
- Organise pairs in different ways on different occasions: familiar maths partners (who can communicate well); pairs of friends (who enjoy working together); children of differing abilities (who can learn something from each other); someone they don't know (to get them used to talking and listening respectfully to any other person).
- Give children working in pairs and groups some time for independent thought and work.
- Support pairs when they prepare to report back to the class. Go over with them what they have done or discovered and what they might say about this. Help them make brief notes – just single words or phrases – to remind them what they are going to say. If you are busy, ask an assistant or another child to take over your role. Then, when it comes to feedback time, support them by gentle probes or questions: "What did you do next?" or "What do your notes say?"

Classroom techniques used in this book

Ways of working

Peer tutoring
pairs of children

good for

This technique can benefit both the child who is being 'taught' and also the 'tutor' who develops a clearer understanding of what they themselves have learned and, in explaining it, can make new connections and solidify old ones. Children often make the best teachers, because they are close to the state of not knowing and can remember what helped them bridge the gap towards understanding.

how to organise it

'Peer tutoring' can work informally – children work in mixed ability pairs, and if one child understands an aspect of the work that the other doesn't, they work together in a tutor/pupil relationship to make sure the understanding is shared by both. Alternatively, you can structure it more formally. Observe children at work and identify those who are confident and accurate with the current piece of mathematics. Give them the title of 'Expert' and ask them to work with individuals needing support. Don't overuse this: the tutor has a right to work and learn at their own level, and tutoring others should only play a small part in their school lives.

Talking partners
pairs of children

good for

This technique helps children develop and practise the skills of collaboration in an unstructured way. Children can articulate their thinking, listen to one another and support each other's learning in a 'safe' situation.

how to organise it

Pairs who have previously worked together (for example, 'One between two', below) work together informally. The children in these pairs have had time to build up trust between them, and should have the confidence to tackle a new, less structured task. If you regularly use 'Talking partners', pairs of children will get used to working together. This helps them develop confidence, but runs the risk that children mutually reinforce their misunderstandings. In this case, changing partners occasionally can bring fresh life to the class by creating new meetings of minds.

One between two
pairs of children

good for

This technique helps children develop their skills of explaining, questioning and listening – behaviours that are linked to positive learning outcomes. Use it when the children have two or more problems or calculations to solve.

how to organise it

Pairs share a pencil (or calculator or other tool), and each assumes one of two roles: 'Solver' or 'Recorder'. (Supplying just one pencil encourages children to stay in role by preventing the Solver from making their own notes.)

The Solver has a problem and works through it out loud. The Recorder keeps a written record of what the Solver is doing. If the Solver needs something written down or a calculation done on the calculator, they must ask the Recorder to do this for them. If the Recorder is not sure of what the Solver is doing, they ask for further explanations, but do not engage in actually solving the problem. After each problem, children swap roles.

Introduce this way of working by modelling it yourself with a confident child partner: you talk through your own method of solving a problem, and the child records this thought process on the board.

Barrier games/Telephone conversations

pairs of children

good for

These techniques help children focus on spoken language rather than gesture or facial expression. The children must listen carefully to what is said, because they cannot see the person speaking.

how to organise it

Barrier games focus on giving and receiving instructions. Pairs of children work with a book or screen between them, so that they cannot see each other's work. The speaker gives information or instructions to the listener. The listener, in turn, asks questions to clarify understanding and gain information.

In 'Telephone conversations', the technique is taken further, as children sit back to back, with only imaginary 'telephones' for conversation.

Rotating roles

groups of various sizes

good for

Working in a small group to solve a problem encourages children to articulate their thinking and support each other's learning.

how to organise it

Careful structuring discourages individuals from taking the lead too often. Assign different roles to the children in the group: Chairperson, Reader, Recorder, Questioner, and so on. Over time, everyone has a turn at each role. You may wish to give children 'role labels' to remind them of their current role.

When you introduce this technique, model the role of chairperson in a group, with the rest of the class watching. Show how to include everyone and then discuss with the children what you have done, so as to make explicit techniques that they can use.

Discussion

Think, pair, share
groups of four

good for

Putting pairs together to work as a group of four helps avoid the situation where children in pairs mutually reinforce their common misunderstandings. It gives children time to think on their own, rehearse their thoughts with a partner and then discuss in a larger group. This encourages everyone to join in and discourages the 'quick thinkers' from dominating a discussion.

how to organise it

The technique is a development of 'Tell your partner' and involves the following:
- One or two minutes for individuals to think about a problem or statement and, possibly, to jot down their initial thoughts
- Two or three minutes where individuals work in pairs to share their thoughts
- Four or five minutes for two pairs to join together and discuss
- If you wish, you can also allow ten minutes for reporting back from some or all groups and whole-class discussion.

You can vary this pattern and the timings, but always aim to give children some 'private' thinking time.

Talking stick
any number of children

good for

Giving all children a turn at speaking and being listened to.

how to organise it

Provide the class with decorated sticks, which confer status on whoever holds them. Then, in a small or large group (or even the whole class), make it the rule that only the person holding the stick may speak, while the other children listen. You can use the stick in various ways: pass it around the circle; tell the child with the stick to pass it to whoever they want to speak next; have a chairperson who decides who will hold the stick next; ask the person with the stick to repeat what the previous person said before adding their own comments or ideas.

Tell your partner
pairs

good for

Whole-class question-and-answer sessions favour the quick and the confident and do not provide time and space for slower thinkers. This technique involves all children in answering questions and in discussion.

how to organise it

Do this in one of two ways:

- When you have just asked a question or presented an idea to think about, ask each child to turn to their neighbour or partner and tell them the answer. They then take turns to speak and to listen.
- Work less formally, simply asking children to talk over their ideas with a partner. Children may find this sharing difficult at first. They may not value talking to another child, preferring to talk to the teacher or not expressing their ideas at all. In this case, do some work on listening skills such as timing 'a minute each way' or repeating back to their partner what they have just said.

Devil's advocate

any number of children

good for

Statements – false or ambiguous as well as true – are often better than questions at provoking discussion.

how to organise it

In discussion with children, take the role of 'Devil's advocate', in which you make statements for them to agree or disagree with and to argue about.

To avoid confusing children by making false statements yourself, mention 'a friend' or 'someone you know' who makes these statements (a version of the 'silly teddy' who, in Nursery and Reception, makes mistakes for the children to correct). Alternatively, explain that when you make statements with your hands behind your back, your fingers may be crossed and you may be saying things that are not true.

Reporting back

Ticket to explain

individuals

good for

This is a way of structuring feedback which helps children get the maximum out of offering explanations to the class. Everyone hears a method explained twice, and children have to listen carefully to their peers, rather than simply think about their own method.

how to organise it

When individuals want to explain their method of working to the class, their 'ticket' to be able to do this is to re-explain the method demonstrated by the child immediately before them. Or children work with a partner and explain their ideas to each other. When called on to speak, they explain their partner's idea and then their own.

Heads or tails
pairs of children

good for

When pairs of children work together, one child may rely heavily on the other to make decisions and to communicate or one child may take over, despite the efforts of the other child to have a say. This technique encourages pairs to work together to understand something and helps prevent an uneven workload.

how to organise it

Invite pairs to the front of the class to explain their ideas or solutions. When they get to the front, ask them to nominate who is heads and who is tails, then toss a coin to decide which of them does the talking. They have one opportunity to 'ask a friend' (probably their partner). As all children in the class know that they may be chosen to speak in this way, because the toss of the coin could make either of them into the 'explainer', they are motivated to work with their partner to reach a common understanding. Assigning the choice of explainer to the toss of a coin stops children feeling that anyone is picking on them personally (do warn them in advance, though!). Variation: If a pair of children has different ideas on a topic, ask both to offer explanations of each other's ideas.

1, 2, 3, 4
groups of four

good for

This technique offers the same benefits as 'Heads or tails', but is used for groups of four children rather than pairs.

how to organise it

This is a technique identical to 'Heads or tails', but with groups of four. Instead of tossing a coin, children are numbered 1 to 4, and the speaker is chosen by the roll of a dice (if 5 or 6 come up, simply roll again).

Additional techniques

Below are some further classroom techniques that are referred to in the lessons in this book.

Ideas map
whole class

good for

This technique enables children to identify what they know and what they don't know and so equips them to monitor their own learning.
Drawing up an ideas map at the start of teaching a topic can produce valuable

material on which to base initial assessments of the children's understanding. It also serves to steer the children towards what they will be learning. Return to the map and even revise and adapt it after a period of teaching in order to consolidate children's learning.

how to organise it

Develop your first ideas maps as a whole class. Hold a creative brainstorm with the children to conjure up as many terms connected to a topic as they can think of, with you adding more terms as appropriate – scribe the contributions and, as you do so, prompt the children to help you establish connections between them.

Once children are used to the idea, they can work in groups or as individuals to draw up ideas maps of their own. They can either work from scratch or you can start them off with a whole-class creative brainstorm, where you collect terms but don't make any connections. Children then devise their own ideas map, using some or all of these terms.

When they construct the map, children aim to link terms together and write on the links something to describe the nature of the connection. Some teachers ask children to write individual words or phrases on sticky notes, so that they can move them around and explore different links before settling on the ones they want to describe.

Ideas board

whole class

good for

An ideas board is a place where children display their work to the rest of the class informally and quickly. It also provides a useful place for you to record ideas and problems that you want children to think about.

how to organise it

The visual aspect of display is not a priority with an ideas board – it is more like a notice board where ideas and information can be shared. Make sure you remove items regularly to keep it fresh and up to date.

Chewing the fat

any number of children

good for

Leaving ideas or questions unresolved provides thoughtful children with the opportunity to extend their thinking and can help develop good habits. Many real mathematicians like to have problems to think about in odd moments, just as some people like crossword clues or chess moves to occupy their mind.

how to organise it

Sometimes end a lesson with ideas, problems or challenges for children to ponder in their own time as you may have run out of time or one of the children has come up with a question or an idea which can only be discussed the next day.

Reframing

any number of children

good for

'Reframing' alters the meaning of something by altering its context or description. It helps children find their way into a difficult or new idea by hearing it rephrased and enlarged.

how to organise it

Rephrase children's words using a variety of language: "You read that out as 'twenty-five multiplied by seven'. That means seven lots of 25." After a few seconds, say: "Imagine a pile of 25 beans, and you've got seven piles like that."

Charts

Classroom techniques

This chart shows which of the classroom techniques previously described are used in which lessons.

	NUMBERS AND THE NUMBER SYSTEM	FRACTIONS, DECIMALS, PERCENTAGES, RATIO AND PROPORTION	ADDITION AND SUBTRACTION	MULTIPLICATION AND DIVISION	HANDLING DATA	MEASURES	SHAPE AND SPACE
	Lesson	Lesson	Lesson	Lesson	Lesson	Lesson	Lesson
One between two		5	12	13, 14		22	26
Talking partners	1		11			21, 24	
Rotating roles	2	6					
Peer tutoring		7	10				
Barrier games / Telephone conversations	4						25, 28
Tell your partner		8				23	
Devil's advocate							27
Think, pair, share	3			16	17, 19		
Ticket to explain				15	18		
Heads or tails / 1, 2, 3, 4			9		20		

Speaking and listening skills

This chart shows which speaking and listening skills are practised in which lessons.

	NUMBERS AND THE NUMBER SYSTEM	FRACTIONS, DECIMALS, PERCENTAGES, RATIO AND PROPORTION	ADDITION AND SUBTRACTION	MULTIPLICATION AND DIVISION	HANDLING DATA	MEASURES	SHAPE AND SPACE
	Lesson	Lesson	Lesson	Lesson	Lesson	Lesson	Lesson
Discuss progress of work						23	
Explain and justify thinking	2	6, 8		13	20	22	
Use precise language to explain ideas or give information	4	5	9				26
Share and discuss ideas and reach consensus	1			14	17	21	
Reach a common understanding with a partner	3	7	10			24	25
Use the processes and language of decision making			11				
Contribute to small-group and whole-class discussion				16	19		27
Listen with sustained concentration			12	15	18		28

Personal skills

This chart shows which personal skills are practised in which lessons.

	NUMBERS AND THE NUMBER SYSTEM	FRACTIONS, DECIMALS, PERCENTAGES, RATIO AND PROPORTION	ADDITION AND SUBTRACTION	MULTIPLICATION AND DIVISION	HANDLING DATA	MEASURES	SHAPE AND SPACE
	Lesson	Lesson	Lesson	Lesson	Lesson	Lesson	Lesson
Organise work							
Plan ways to solve a problem					17	24	
Plan and manage a group task			11				
Work on a task with several aspects						23	
Use different approaches to tackle a problem				15			
Organise findings	3						
Work with others							
Discuss and agree ways of working					18		
Work cooperatively with others	1	5		14	19	21	
Overcome difficulties and recover from mistakes		6					26
Show awareness and understanding of others' needs	4	7					28
Improve learning and performance							
Reflect on learning			10				
Critically evaluate own work			12				25
Assess learning progress				13		22	
Take pride in work			9				
Develop confidence in own judgements	2	8			20		27

Lessons

Numbers and the number system

Learning objectives

	Lessons			
	1	**2**	**3**	**4**
ⓜ Maths objectives				
make general statements about the products of odd and even numbers	●			
know and apply simple rules of divisibility		●		
make a general statement about triangular numbers			●	
find the difference between a positive and negative integer				●
ⓢ Speaking and listening skills				
share and discuss ideas and reach consensus	●			
explain and justify thinking		●		
reach a common understanding with a partner			●	
use precise language to explain ideas or give information				●
ⓟ Personal skills				
work with others: work cooperatively with others	●			
improve learning and performance: develop confidence in own judgements		●		
organise work: organise findings			●	
work with others: show awareness and understanding of others' needs				●

About these lessons

Lesson 1: Odd and even number products

 Make general statements about the products of odd and even numbers

In this activity, children explore the results of multiplying odd and even numbers to find that the product of two even numbers is even; the product of two odd numbers is odd; and the product of an odd and an even number is even.

 Share and discuss ideas and reach consensus

Classroom technique: Talking partners

Children discuss and agree together the general statements about the results of multiplying odd and even numbers.

 Work with others: work cooperatively with others

Pairs find their own way to work together on a task, but are reminded to share the work so that both children are equally engaged in problem solving.

Lesson 2: Exploring divisibility

 Know and apply simple rules of divisibility

Tests of divisibility are useful aids for mental calculations, as they may indicate a 'way in' to a calculation that appears difficult. In this lesson, children practise the different tests of divisibility and discover what they are already familiar with and what they need to tackle further.

 Explain and justify thinking

Classroom technique: Rotating roles

Children work in groups of three to decide whether statements about multiples, factors and divisibility are true or false. Each time, two of them discuss the statement, with the third child summing up their decision. Children take on each role in turn during the activity.

 Improve learning and performance: develop confidence in own judgements

Children discuss statements about divisibility which may be true or not. This helps them become clearer about their own judgement about the truth of the statement.

Lesson 3: Triangular numbers

 Make a general statement about triangular numbers

Working with number sequences helps children see that pattern is an important feature of mathematics. In this activity, children generate the sequence of triangular numbers, identify the pattern and explore what happens when consecutive triangular numbers are added.

 Reach a common understanding with a partner

Classroom technique: Think, pair, share

Children work as individuals on the problem about consecutive triangular numbers, then discuss their ideas with their partner. Finally, pairs combine forces with another pair to discuss their work, listen to each other's ideas and agree a statement.

 Organise work: organise findings

After completing their exploration, children work together to prepare a report on what they have discovered about triangular numbers.

Lesson 4: Positive and negative numbers

 Find the difference between a positive and negative integer

Negative numbers can appear very abstract. In this activity, a timeline showing events before and after a child is born provides a realistic context and helps children visualise how negative and positive numbers work.

 Use precise language to explain ideas or give information

Classroom technique: Telephone conversation

Each child constructs their own timeline, then sits back to back with their partner and instructs them how to create one just the same.

Work with others: show awareness and understanding of others' needs

The child who is describing how to construct a timeline like theirs must put themselves in their partner's shoes to decide how to phrase their information and how to pace it.

Odd and even number products
Classroom technique: Talking partners

Maths
Make general statements about the products of odd and even numbers

Speaking and listening
'Share ideas and reach agreement'
Share and discuss ideas and reach consensus

Personal skills
'Work cooperatively with others'
Work with others: work cooperatively with others

Words and phrases
odd, even, total, multiply, product, tree diagram, convince, prove, check, test, general statement

Resources
for each pair:
copy of RS1
digit cards (0–9)
slips of paper

Generalising
This is an opportunity to revise what children already know about odd and even numbers and to practise making generalisations.
Children should know the rules:
$O +/- O = E$
$O +/- E = O$
$E +/- E = E$

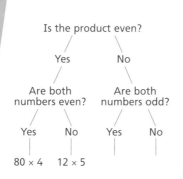

Making general statements
For example, children might write, "When you multiply two odd numbers, the answer is always odd." They may also use symbols: $O \times E = E$. Children check their general statements with different examples.

Introduction

With the children's help, write up a dozen addition and subtraction equations. Children work in pairs to agree general statements about the results (in terms of odd and even) of adding and subtracting odd and even numbers.

Collect in and agree the children's ideas and ask for explanations.

(m) *If I subtract 37 from 83, will the answer be odd or even? Why do you think that?*

(m) *If I add several numbers together, what combinations of odd and even give me an even total?*

(☺) *Does adding three odd numbers always give an odd total? Even if the numbers are really big? Why is that?*

Pairs

Give each pair of children a copy of RS1 and a set of digit cards. Pairs choose three cards, arranging them to make a multiplication: ▢▢ × ▢ or ▢ × ▢▢.
Children write the equation on a slip of paper and agree the product. They then move the number sentence along the tree on RS1 and write the multiplication at the end of the correct 'branch'.

After a couple of minutes, stop the class and check that children understand how to use, and record on, the tree diagram. Make sure children are sharing the work equally. Remind them, if necessary, that they also need to discuss and share the decision making.

Pairs agree and write down a general statement (in terms of odd and even) about the sets of calculations at the end of each of the four branches. Ask pairs to make sure that both partners understand the statements and are prepared to explain them to the class.

(m) *Do the branches have roughly the same number of answers? Why?*

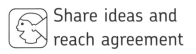

With your partner, find a multiplication calculation for each branch.

What do you do when you don't agree about something?

Support: Use two digit cards to make a multiplication and a multiplication grid to find the answers.

Extend: Explore whether there are any general statements to make about division involving odd and even numbers. Alternatively, explore multiplying three numbers.

Plenary

Take feedback from one or two pairs. As a class, agree a clear statement for each branch and write this up, using symbols.

> Multiplying two even numbers gives you an even product:
>
> E × E = E
>
> Multiplying an odd and an even number gives you an even product:
>
> E × 0 = E

Without multiplying 223 and 534, what can you say about their product? Why is it helpful to have this knowledge?

Discuss with your neighbour whether this statement can be expressed more clearly.

Assessment for learning

Can the children	**If not**
Recognise whether a multiplication calculation will have an odd or an even product and say how they know?	Model the multiplication of small numbers on squared paper and demonstrate how one even number ensures an even product.
Readily turn to their partner and discuss an idea in pairs?	Ask each child to write a draft of a statement and pass it to their partner to look at and comment on. Help the class develop a habit of paired discussion, using 'Tell your partner' (p10).
Share the work of recording on RS1?	Remind children that they need to practise and develop all their skills.

Exploring divisibility
Classroom technique: Rotating roles

Tests of divisibility

3 The sum of its digits is a multiple of 3.

6 The sum of its digits is a multiple of 3 and it is even.

10 The last digit is a 0.

5 The last digit is a 0 or 5.

2 It is even.

4 The last two digits are divisible by 4 OR halve it twice and get a whole number.

8 Halve it three times and get a whole number.

9 The sum of its digits is a multiple of 9.

Prime numbers are only divisible by themselves and 1. (1 is not a prime number.)

The recorder's roles

The child who is recording needs to listen carefully to their partners' discussion and suggest a sentence to record, checking this with their partners.
If necessary, model how to work this way, working with two children on an invented statement of your own.

Calculators

Once they have agreed their statement, Child A can check if it is true with a calculator.

Introduction

Write up these properties and numbers:

divisible by 3	multiple of 4
prime number	divisible by 5
multiple of 9	divisible by 8
has a factor of 10	divisible by 6
57 162 75	84 96 400

Children choose a number that fits each property (numbers can be used more than once) and agree how they know this.

Take feedback, establish each rule of divisibility in turn and write it on the board.

(m) *Does this statement fit more than one number?*

Tell your neighbour why 400 is a multiple of so many different numbers.

Tell your neighbour which of these properties you are really sure about and which you aren't.

Groups of three

Give each group of three children a set of statement cards cut from RS2.

Child A reads out one of their statement cards. Child A and Child B discuss whether or not they think the statement is true and establish a reason. Child C records the reason in one sentence.

Children swap roles after each statement.

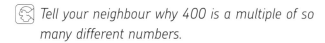

171 is a multiple of 9 because 1 and 7 and 1 makes 9.

Can 41 be divided by 3? How do you know?

Are you sure that an even number can't be a prime number? Is that always true?

Explain why this number is a multiple of 9.

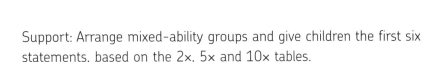
Support: Arrange mixed-ability groups and give children the first six statements, based on the 2×, 5× and 10× tables.

Extend: Each child makes up one or two true or false statements, using three-digit or bigger numbers. The group work with these in the same way as with the given statements.

Plenary

Display one of the children's explanations, hiding the actual number, Pairs of children need to think of a two-, three- or four-digit number to fit. Collect in some suggestions.

Repeat this with other explanations.

Finally, display RS2 and read it out with the children. Children identify to their partner some of the statements they feel confident about.

How could the four-digit number 473☐ be completed to make it a multiple of 9?

If a number can't be divided by 8, 9, 6 or 5, it is a prime number. True or false?

Which rule are you most sure about?

Which rule for divisibility by 8 do you prefer? Why?

Assessment for learning

Can the children

Recognise whether a number is divisible by 2, 3, 4, 5, 6, 8, 9 or 10?

Explain how they know that a number is, or is not, divisible by a particular number?

Confidently state that they understand and can use one or more tests of divisibility?

If not

Work with children to examine multiples from, for example, the 5 or 6 multiplication tables and agree what they all have in common. Use this to check for divisibility of slightly larger numbers.

Give two different explanations yourself, one clear, the other muddled, and ask the children to discuss which one is better, and why.

Give children a sheet showing all the tests for divisibility to tick those they know and put a question mark by those they are not sure about. Provide opportunities to practise using the 'wobbly' ones.

Triangular numbers

Classroom technique: Think, pair, share

Learning objectives

Maths
Make a general statement about triangular numbers

Speaking and listening
'Reach an understanding with your partner'
Reach a common understanding with a partner

Personal skills
'Organise your results'
Organise work: organise findings

Words and phrases
triangular number, square number, continue, consecutive, adjacent, sequence, pattern, predict, check, general statement

Resources
display squared grid squared paper

Introduction

Display a squared grid and demonstrate how to illustrate triangular numbers on the squares. Children draw the first four triangular numbers on their squared paper.

Collect in and record the sequence of triangular numbers and the difference between each adjacent pair.

triangular number 1   3   6   10

difference   2   3   4

m How do the triangular numbers grow?

m What would the next two numbers be?

👤 Do you all understand what Maya has just said? Can you repeat it?

Pairs/Groups of four

Children first work in pairs, drawing and recording the first ten triangular numbers.

Pairs then join in groups of four and explore what happens when they add any pair of adjacent numbers. Once they have agreed on the rules, they prepare a report on what they have discovered about triangular numbers.

👤 Tell your partner what to do to get the next triangular number.

👤 Was there anything you didn't understand in that explanation?

😊 How can you record the numbers clearly to show how the pattern grows?

Support: Get children to work neatly and to check their work carefully. The visual aspect of this work should provide support, as long as children avoid errors.

Extend: Children find the first 20 triangular numbers and check that the rules apply with the higher numbers.

Think, pair, share
Children work on this as individuals for a few minutes, then discuss their ideas with their partner. Finally, pairs form groups of four to discuss their work, listen to each other's ideas and agree a statement.

What happens?
Adding two consecutive triangular numbers gives a square number – for example, $3 + 6 = 9$ and $15 + 21 = 36$.

Reach an understanding
with your partner

Completing work

Give children time to complete their reports and display these so children can read each other's. This encourages them to take care about, and pride in, their work.

Plenary

The class record the triangular numbers, the differences between them and the totals of consecutive numbers. Establish that these totals form the sequence of square numbers. Occasionally, make a deliberate error and ask children to spot it when you do so.

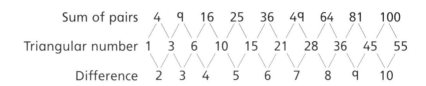

Sum of pairs		4	9	16	25	36	49	64	81	100
Triangular number	1	3	6	10	15	21	28	36	45	55
Difference		2	3	4	5	6	7	8	9	10

ⓜ *Is there anything you want to change in this explanation of how triangular numbers grow?*

ⓜ *How do you know that the number I just wrote was wrong?*

🗨 *In which way did talking together as a group of four help your thinking?*

Assessment for learning

Can the children

ⓜ Find what comes next in the sequence of triangular numbers?

🗨 Talk together about their work, rather than one child working on it and the other idling?

☺ Organise their report clearly?

If not

ⓜ Encourage children to work tidily as this makes it easier to count squares accurately. If a child's addition skills are poor, help them use strategies for adding one- and two-digit numbers.

🗨 Consider putting children with different partners. Use the 'One between two' technique (p8) which requires children to cooperate and talk about their work.

☺ Arrange pairs of children where one is an efficient and organised recorder who can teach their skills to their partner.

Positive and negative numbers
Classroom technique: Telephone conversation

Introduction

Display RS3 and a squared grid. Children construct a personal timeline, with zero indicating when the child was born.

Introduce the year of the child's birth as Year 0 and work with the class to place each event on the timeline.

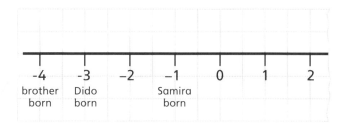

Negative numbers
Encourage the children to say 'negative' when referring to positions on the line: "Dido was born at negative 3."

m *If this is zero and each step is worth 1, what is the value of this mark?*

How many steps is it from negative 4 to positive 2? How do you work that out?

Pairs

Each child now makes their own timeline, showing their own events.

Preparing for the lesson
Before this lesson (and if appropriate), ask children to find out dates of events that happened since they were born and up to 10 years before they were born. Otherwise, work with imaginary events or characters from fiction.

Children sit back to back and construct their timeline in private. When they have finished, they work as pairs. Child A has their own timeline in front of them and instructs Child B to draw a timeline showing the same events. They then swap roles to replicate Child B's line.

At the end, they compare timelines to check that they are identical.

What is the first thing you need to tell your partner?

How would you like your partner to behave when they are doing the talking and you are drawing the line?

What does your partner need from you?

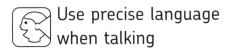
Support: Provide ready-drawn timelines, with no events on.

Extend: Children think about how they might record a simple calculation involving negative numbers.

Plenary

With the help of the class, quickly sketch a new timeline, showing events for a composite child (made up of events that children put on their own timelines).

Ask some questions about this for pairs to discuss. Pairs then invent one or two questions of their own to ask the class – for example, "How old was Charlie when they started school?"; "Which two events were six years apart?"

 Can you work out the distance from negative 3 to positive 4 in your head?

How did you work out that answer?

Assessment for learning

Can the children

Find the difference between positive and negative integers using a number line? Working mentally?

Describe to a partner how to place the numbers on their line?

Speak clearly and avoid showing impatience?

If not

Do more work on negative numbers in contexts such as temperature or a lift travelling above/ below ground.

Get a child to ask questions for their partner to answer.

Make sure to model these behaviours yourself and to comment favourably on them when displayed by children.

Name _____

Self and peer assessment

Lesson 1: Odd and even number products	I think	My partner thinks
(m) When I multiply one even and one odd number, I know whether the product will be odd or even.		
I talk with my partner about our work until we agree what to do.		
I share the work with my partner and take turns.		

Lesson 2: Exploring divisibility	I think	My partner thinks
(m) I can tell when a number is divisible by 2, 3, 4, 5, 6, 8, 9, 10.		
I can explain to my partner how I know a number is, or is not, divisible by 4.		
I know when I am sure about a divisibility statement and when not.		

Name _____

Lesson 3: Triangular numbers	I think	My partner thinks
(m) I can find what comes next in a sequence of triangular numbers.		
I talk with my partner about our work so that we both understand it.		
I organise my report clearly so people will understand it.		

Lesson 4: Positive and negative numbers	I think	My partner thinks
(m) I can find the difference between a positive and a negative number using a number line or working mentally.		
I can describe to my partner where to place the numbers on the line.		
I speak clearly when we are working back to back.		

Self and peer assessment

Fractions, decimals, percentages, ratio and proportion

Learning objectives

	Lessons			
	5	6	7	8
ⓜ Maths objectives				
convert improper fractions to the equivalent mixed numbers	●			
convert fractions to decimals		●		
use fractions as an operator to find fractions of an amount			●	
round a number with two decimal places to the nearest tenth				●
ⓢ Speaking and listening skills				
use precise language to explain ideas or give information	●			
explain and justify thinking		●		●
reach a common understanding with a partner			●	
ⓟ Personal skills				
work with others: work cooperatively with others	●			
work with others: overcome difficulties and recover from mistakes		●		
work with others: show awareness and understanding of others' needs			●	
improve learning and performance: develop confidence in own judgements				●

About these lessons

Lesson 5: From fractions to mixed numbers

(m) Convert improper fractions to the equivalent mixed numbers

Children play a game where they convert improper fractions to mixed numbers and find the position of these numbers on a number line.

Use precise language to explain ideas or give information

Classroom technique: One between two

Children have to tell their partner where a number goes on the number line by giving precise instructions. The listener may ask questions to clarify the instructions.

Work with others: work cooperatively with others

The success of this activity depends on pairs working together and cooperating. In the plenary, the class discuss how to make working with a partner a pleasant experience.

Lesson 6: From fractions to decimals

(m) Convert fractions to decimals

Children place improper fractions and mixed numbers in order. They use a calculator to convert the fractions to decimal numbers to help them compare their values.

Explain and justify thinking

Classroom technique: Rotating roles

Each child has a different role to play and, during the activity, has experience of all three roles. When they place a fraction on the 'ladder', children have to justify their decision.

Work with others: overcome difficulties and recover from mistakes

With this activity, the focus is not on getting correct answers but on making choices – with an outcome that depends partly on chance. This frees children from struggling with the problems they encounter and allows them to learn from any mistakes they make, rather than worry about making errors.

Lesson 7: Fractions of an amount

(m) Use fractions as an operator to find fractions of an amount

Children deal with word problems where they must divide and multiply to find fractions of various amounts – for example, to find $\frac{4}{5}$ of an amount, they have to find $\frac{1}{5}$ by dividing, then multiply the result by 3.

Reach a common understanding with a partner

Classroom technique: Peer tutoring

Children work individually on the questions before discussing their calculations and their choices with a partner. They tutor each other when either of them has made an error, which supports the understanding of both the tutor and the pupil.

Work with others: show awareness and understanding of others' needs

Tutoring a partner requires children to think about what that partner does, and does not, understand, and how to support them sensitively.

Lesson 8: Rounding decimal numbers

(m) Round a number with two decimal places to the nearest tenth

Children practise rounding decimal numbers on a number line in the context of a game. The game gives them the incentive to think about the rules of rounding and to aim for accuracy. The number line provides a model for children to use when rounding numbers mentally.

Explain and justify thinking

Classroom technique: Tell your partner

Both children choose a number and round it, then tell their partner about this, with the aim of choosing one of the numbers to use in the game. This involves children in giving reasons for their choice and in listening to their partner.

Improve learning and performance: develop confidence in own judgements

As children earn points for their team with the decimal numbers they round up or down, they gain confidence in their ability to work with decimal numbers as well as in their own decision making.

From fractions to mixed number

Classroom technique: One between two

Introduction

Write up the digits 4, 3 and 2. Children work in pairs to make improper fractions, using these digits. Collect several such numbers on the board.

$$\frac{32}{4} \qquad \frac{44}{3} \qquad \frac{23}{3} \qquad \frac{33}{3} \qquad \frac{23}{2}$$

Pairs discuss how to change improper fractions to mixed numbers, aiming to agree an explanation. Ask some pairs for their ideas and agree a class explanation.

(m) *Why didn't I write up the digits 1, 2 and 3 for you to use? What happens when the denominator is 1?*

(m) *Why isn't $\frac{4}{32}$ an improper fraction?*

(s) *Explain how you know which of these numbers are less than 10.*

Teaching the game
Demonstrate the activity with one child in front of the class and model how to give precise instructions: "$4\frac{2}{3}$ is more than $4\frac{1}{4}$. It goes two thirds of the way from 4 to 5." Emphasise that the child doing the writing must check that the mixed number and its place on the number line are correct.

Pairs

Child A rolls the three 1–6 dice and makes an improper fraction less than 10, using the digits. Child B jots down the fraction on a sheet of paper. Child A then works out loud to change this fraction to a mixed number and tells Child B where to write this (as a mixed number and an improper fraction) on the first number line on RS4, together with an initial.

Children then swap roles and continue taking turns until one of them has three numbers on the number line with none of their opponent's in between.

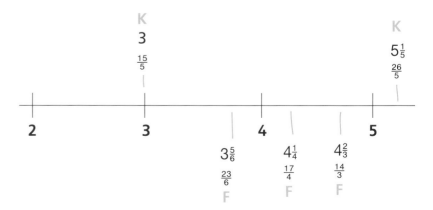

Fiona wins this game as she has three fractions next to each other.

(m) *What fraction might go between the two numbers your partner has made?*

(👤) *How can you explain where $\frac{26}{8}$ goes on the number line?*

(😊) *What do you want your partner to do to help you? And how can you be helpful to your partner?*

Support: Use the 0–10 line marked with quarters and halves on RS4, 10–20 number cards (cut from RS5) and one dice marked with just 2 and 4. Children take a card and roll the dice to make the improper fraction.

Extend: Use 0–9 dice.

Plenary

Write up a challenge:

Solution
Arrange the digits thus: $\frac{13}{4} = 3\frac{1}{4}$

> **Put each digit in two boxes to make the number sentence correct.**
>
> 1   3   4
>
> $$\frac{\Box\Box}{\Box} = \Box\frac{\Box}{\Box}$$

Pairs discuss how to solve this. Then establish a solution with the class.

Finally, ask for ideas about how to make working with a partner a pleasant experience – or the opposite.

Assessment for learning

Can the children

(m) Convert an improper fraction to an equivalent mixed number?

(👤) Explain where fractions should be placed on the number line?

(😊) Identify and display one piece of behaviour that facilitates working with a partner?

If not

(m) Work on modelling the process: cut up whole sheets of paper into quarters and write the matching improper fraction on several of these. Ask the child how many whole sheets of paper and how many quarters this equals and show that number on the number line.

(👤) Model this to children, explaining to them where to place numbers on a number line marked with whole numbers and halves.

(😊) Talk with the class about 'helpful' and 'unhelpful' behaviours and encourage children to give positive feedback when their partners cooperate well.

From fractions to decimals

Classroom technique: Rotating roles

Dividing the numerator
If children have done plenty of work on the idea that fractions are about dividing (for example, sharing 7 pizzas between 5 people), then dividing the numerator by the denominator should make sense to them.

Collaborating
Any member of the group may suggest or ask questions, but Child C has main responsibility for checking they all agree with Child B's decision.

Introduction

Display these fractions:

$$\frac{2}{3} \quad 1\frac{5}{10} \quad \frac{5}{8} \quad 2\frac{3}{4} \quad \frac{13}{5} \quad \frac{12}{4} \quad \frac{13}{10} \quad \frac{3}{4} \quad 1\frac{1}{2}$$

Pairs work together to place the numbers on a sketched 0–3 number line. Explain that you don't expect great accuracy, but that the numbers should be in the correct order.

Children show where each number would be placed on the number line.

Demonstrate using a calculator to convert fractions to decimals to check the positions of the numbers – for example, with $2\frac{3}{4}$ keep the 2 as it is and divide the numerator (3) by the denominator (4) to get 2.75. Alternatively, convert $2\frac{3}{4}$ to an improper fraction and divide the numerator (3) by the denominator (4) to get 2.75.

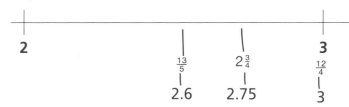

(m) *Which of these fractions are bigger than 1?*

(speaking) *Are any of the numbers not in the correct position? How do you know that?*

(personal) *Were any numbers hard to place on the number line? Why were they difficult?*

Groups of three

Working in groups of three, children place the cards cut from RS6 face down on the table. Child A picks a card and reads the number. Child B writes the number on the ladder, deciding with Child A where exactly it should go. Child C asks questions to clarify Child B's reasoning. Child C may use the calculator if there is disagreement.

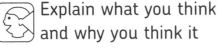
The group rotate roles after each number.

When a ladder is complete, the group cross out any numbers which are badly placed and work out how many are now in order.

Why does dividing 7 by 5 give you the decimal equivalent of $\frac{7}{5}$?

How did you decide where to place the first fraction?

Show me how you could swap the fractions round so they are in order.

Support: Use cards cut from RS7.

Extend: Once children have a set of fractions in order on the ladder, they agree fractions or decimals that go between them.

Plenary

Display RS8. Groups choose any fraction between 0 and 2 and work as a class to place these in order on the ladder. Occasionally, make a deliberate error (see 'Devil's advocate', p11) and ask children to explain in what way you are mistaken.

Convince me that $1\frac{4}{8}$ is greater than $1\frac{2}{5}$.

I made an error with that fraction. Thank you for helping me get it right.

Getting over difficulties
Emphasise that the activity is not about getting correct answers but making choices. Children will only get all six numbers in order if they are very lucky.

Assessment for learning

Can the children

Convert several fractions to decimals, then put them in order?

Explain how they know a fraction is greater or less than another one?

Acknowledge any difficulties they had or errors they made?

If not

Do some work focusing on the idea of fractions as numbers divided by other numbers (3 pizzas shared by 4 children, 2 apples shared between 5 children) and represent the results as fractions and as decimals.

Partner children who can do this with less confident ones. Ask the first child to offer an explanation and the second child to repeat this back (this is a version of 'Peer tutoring', p8).

Model the making of errors yourself to demonstrate that it is a normal part of mathematical work. Emphasise the value of checking one's own work to spot and correct errors.

Fractions of an amount
Classroom technique: Peer tutoring

Learning objectives

Ⓜ Maths
Use fractions as an operator to find fractions of an amount

Speaking and listening
'Reach an understanding with your partner'
Reach a common understanding with a partner

Personal skills
'Think about what other people need'
Work with others: show awareness and understanding of others' needs

Ⓦ Words and phrases
fraction, numerator, denominator, equivalent fraction, divide, multiply, explain, calculate, check

Ⓡ Resources
display copy of RS9
for each pair:
wipe boards
copy of RS10

Relate fractions to division
Emphasise the links with division and multiplication. To find $\frac{1}{10}$, you divide by 10. To find $\frac{3}{10}$, you multiply the tenth by 3. Make sure children are reading and saying the fraction names correctly.

Peer tutoring
Partner weak children with a competent peer. With these pairs, the tutoring is likely to be one-way. Other pairs can be of similar ability, where tutoring will be more a matter of jointly working together to spot, understand and correct errors.

Introduction

Display the top half of RS9. Discuss the questions as a class.

> Imagine you are in a team, planning to run 15 km for charity.
>
> There are 10 of you in the team, and you all want to run the same amount.
>
> 1. What fraction of the run will you do?
> 2. What distance will you run?
> 3. How do you calculate one tenth of a number?

Establish that children will run one tenth of 15 km (that is, 1.5 km). To calculate this, they divide 15 km by 10.

Display the remaining questions on RS9.

Children discuss these questions in pairs and make notes, if necessary, on a wipe board.

> 4. Suppose some members of your team drop out. You agree to run $\frac{3}{10}$ of the run. How far will you run?
> 5. How did you work it out?
> 6. Write a number sentence to show this working.

Take feedback and agree suitable calculations to record on the board.

> 15 km ÷ 10 = 1.5 km
> 1.5 km × 3 = 4.5 km

Ⓜ *How does knowing $\frac{1}{10}$ of 15 km help you work out $\frac{3}{10}$ of 15 km? How would you work out $\frac{7}{10}$ of 15 km?*

Did you and your partner agree on the easiest way of working out $\frac{2}{5}$ of 15 km?

Pairs

Children work alone on the first few questions of RS10, thinking about which fraction they would choose each time. After about 10 minutes, they work with a partner to compare and check their calculations and peer tutor each other where there are errors. Pairs then work on the remaining questions on RS10. Provide correct answers, as you see fit.

Reach an understanding with your partner

(m) *What is wrong with that calculation?*

(face) *Can you explain to your partner which bit you don't understand?*

(face) *Can you see what help your partner needs?*

Support: Children work on those questions they feel comfortable with. (The first few questions on RS10 are the easiest.) Make sure children have confident partners.

Extend: Pairs of children work together to create their own questions and answers, similar to those on RS10.

Answers for RS10

A. $\frac{1}{2}$ of £18 is £9.
    $\frac{1}{4}$ of £40 is £10.

B. $\frac{1}{3}$ of 15 pages is 5 pages.
    $\frac{3}{4}$ of 8 pages is 6 pages.

C. $\frac{3}{4}$ of 400 ml is 300 ml.
    $\frac{3}{10}$ of 1000 ml is 300 ml.

D. $\frac{1}{2}$ of 500 m is 250 m.
    $\frac{3}{10}$ of 800 m is 240 m.

E. $\frac{2}{5}$ of £45 is £18.
    $\frac{3}{5}$ of £35.50 is £21.30.

F. $\frac{1}{6}$ of 240 books is 40 books.
    $\frac{3}{4}$ of 52 books is 39 books.

G. $\frac{2}{3}$ of 51 pages is 28 pages.
    $\frac{2}{9}$ of 90 pages is 20 pages

H. $\frac{6}{8}$ of an 880 ml is 660 ml.
    $\frac{3}{10}$ of 1050 ml is 315 ml.

I. $\frac{4}{5}$ of 25 km is 20 km.
    $\frac{5}{8}$ of 32 km is 20 km.

J. $\frac{3}{4}$ of 1 kg and $\frac{2}{3}$ of 1.5 kg come to 1.75 kg.
    $\frac{3}{8}$ of 4 kg is 1.5 kg.

Plenary

Children explain some of their choices.

They then think in silence for a minute about the experience of working as a pair before telling their partner one thing they appreciated about working with them in this lesson.

(m) *In answering these questions, when did you need to multiply?*

(faces) *Agree with your partner a statement about how to use division to find a fraction of an amount.*

(face) *What did you need from your partner? And what do you think they needed from you?*

Assessment for learning

Can the children

(m) Find a fraction of a number or an amount?

(faces) Work with a partner to check a calculation and agree whether it is correct?

(face) Identify and display behaviour that supports their partner?

If not

(m) Model the process with lengths of ribbon – for example, find a 60 cm length by folding and cutting and record this as 60 cm ÷ 4 = 15 cm. Work out the calculation for the three of these quarter lengths: 15 cm × 3 = 45 cm.

(faces) Model this with a child or another adult. Do a simple calculation incorrectly on the board. Get your partner to help you 'understand' your mistake by asking questions. It is important for children to see their teacher making mistakes and being corrected by another.

(face) Brainstorm a class list of 'What we need from our partners' to display and refer to in future lessons.

Rounding decimal numbers

Classroom technique: Tell your partner

Target numbers		
0.1	0.5	0.9
0	0.6	0.8
0.2	0.3	0.7

Rules for rounding
Take the opportunity to establish the rules for rounding numbers with two decimal places to the nearest tenth. Explain that a 5 in the hundredths place rounds up rather than down: 0.35 is rounded up to 0.4. Numbers below 0.05 round down to zero.

Introduction

Display RS11. Start playing a class game on rounding to the nearest tenth. Divide the class in half and give each team a set of three target numbers from the list shown.

Roll three 0–9 dice. Individuals choose two of the digits and create a decimal number of the form 0.☐☐. (They are aiming for a number that rounds to one of their targets.)

Children then 'tell their partner' which number they have made and which 'tenths' number it is nearest to.

Invite a member of each team to give you their chosen number, locate these on the line with a pointer and establish what tenths number each rounds to.

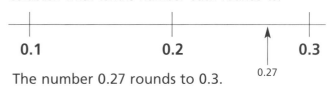

The number 0.27 rounds to 0.3.

If a team has that as a target number, they score a point. Repeat this until both teams have scored a few points.

m *Tell me a decimal number you don't want to make, because it would suit the other team.*

m *Are you sure 0.74 is closer to 0.7 than to 0.8? How can we use the line to check?*

Pat yourself on the back if you chose a number that scored a point for your team.

Groups of four

Groups of four play the same game in teams of two. Children roll three 0–9 dice to generate three digits and record these.

Individuals choose two of the digits to create a decimal number as before, then 'tell their partner' which number they have made and which 'tenths' number it is nearest to. The pair agrees which of their numbers to use. Both pairs locate their number on the number line and show the other pair what it rounds to and whether they score a point.

The game is over when one pair has scored 10 points.

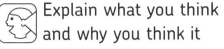
 What is the rule for rounding a number with 5 in the hundredths place?

Why did you choose to make that decimal number?

Are you getting quite good at this game?

Support: Partner less confident children with those who are confident with decimals.

Extend: Children adapt the game to involve decimal numbers with thousandths – for example, 0.026.

More digits

With 9.2 and 34.2, children need to focus on the digits after the decimal point – those to the left of it are not involved in the rounding.

Working mentally

Encourage children to work mentally, but help them sketch number lines where this will further their understanding.

Plenary

Write up these numbers:

| 0.6 | 9.2 | 34.2 |

Pairs choose one of the numbers and create four decimal numbers, three of which round to their chosen number and one that doesn't.

Pairs present their set of numbers and ask the class to spot the odd one out and to explain why it does not round to the chosen number.

 Convince me that 0.65 does not round to 0.6.

Tell you partner how successful you think they are at rounding decimal numbers.

Assessment for learning

Can the children

Mentally round 0.23, 1.55 or 37.09 to the nearest tenth?

Explain why they chose to make a particular number in terms of what it rounds to?

Readily offer a number that does, or does not, round to 0.4? To 13.5?

If not

Practise rounding whole numbers and numbers with one decimal point. Keep using the number line model and getting children to place decimal numbers in roughly the correct position on it.

Offer an explanation of your own for them to assent to: "I expect you chose 2.34 because it rounds to 2.3." Ask children to show you the number's position on a number line.

Introduce other games involving maths at a more suitable level if children are not confident with the maths and encourage them to make statements about their success at these.

Name _____

Self and peer assessment

Lesson 5: From fractions to mixed numbers	I think	My partner thinks
(m) I can convert an improper fraction to a mixed number.		
I can explain where to place a fraction on the number line.		
I try to help my partner so they enjoy working with me.		

Lesson 6: From fractions to decimals	I think	My partner thinks
(m) I can convert fractions to decimals.		
I can explain how I know which of two fractions is larger.		
I keep trying if I have problems with my work or if it is hard.		

Name _____

Lesson 7: Fractions of an amount	I think	My partner thinks
(m) I can find fractions of numbers and amounts.		
(☺) I work with my partner to check our calculations.		
(☺) I try to think what it is like for my partner to work with me.		

Lesson 8: Rounding decimal numbers	I think	My partner thinks
(m) I can round numbers to the nearest tenth.		
(☺) I can explain the rule for rounding a number such as 0.67 to the nearest tenth.		
(☺) I feel confident about choosing good numbers when we play the number game.		

Self and peer assessment

Addition and subtraction

Learning objectives

	Lessons			
	9	10	11	12
Ⓜ Maths objectives				
find the difference between a pair of four-digit numbers	●			
add decimal fractions		●		
solve problems involving money			●	
use inverse operations to solve number puzzles				●
Speaking and listening skills				
use precise language to explain ideas or give information	●			
reach a common understanding with a partner		●		
use the processes and language of decision making			●	
listen with sustained concentration				●
Personal skills				
improve learning and performance: take pride in work	●			
improve learning and performance: reflect on learning		●		
organise work: plan and manage a group task			●	
improve learning and performance: critically evaluate own work				●

Learning objectives

About these lessons

Lesson 9: Finding differences

(m) Find the difference between a pair of four-digit numbers

In this puzzle, children need to find the difference between pairs of numbers to choose two with a difference close to 1000. Rounding the numbers will help children find an approximate difference.

Use precise language to explain ideas or give information

Classroom technique: Heads or tails

Pairs of children are invited to explain their ideas to the class. They toss a coin to decide who will give the explanation. This technique helps ensure that both children agree on a clear explanation and are able to pass it on to the class.

Improve learning and performance: take pride in work

Tackling a puzzle successfully is enjoyable for children and can be a source of pride and confidence in their achievement.

Lesson 10: Adding decimal fractions

(m) Add decimal fractions

When calculating, children should use mental methods as a first resort, aided by jottings where appropriate. In this activity, children work on a problem involving addition of decimal numbers and are encouraged to work mentally as far as possible.

Reach a common understanding with a partner

Classroom technique: Peer tutoring

A child who is weak at calculating with decimals may nonetheless be a competent problem solver. Putting such a child with a partner who is more confident with decimals means that both can work at the problem, with the more confident child acting as tutor and explaining how to add the decimal numbers.

Improve learning and performance: reflect on learning

Encouraging children to reflect on what they have been learning, as they do in this lesson, helps children become involved in, and take charge of, their own development as mathematicians and as individuals.

Lesson 11: Calculating with money

(m) Solve problems involving money

Children are asked to solve a 'real-life' problem about what items to include in a toolkit for Y7 pupils. This involves understanding the task, deciding how to tackle it, estimating answers and working out totals of money accurately.

Use the processes and language of decision making

Classroom technique: Talking partners

Children work in groups of three, discussing the task and sharing the decision making – both about how to manage the task and what items to include in the kits.

Organise work: plan and manage a group task

Sometimes, it is useful to structure group work (using, for example, 'Rotating roles', p9), but children also need experience of organising and sharing out their own work, as in this activity.

Lesson 12: Inverse operations

(m) Use inverse operations to solve number puzzles

Children sometimes have difficulty solving 'box' calculations, where a number is missing, as in $\square \times 3 = 126$. This activity gives children practice in turning such a calculation into inverse number sentences to find the answers.

Listen with sustained concentration

Classroom technique: One between two

Children invent and solve number puzzles together with a partner. The child with the pencil writes only what they are instructed to by their partner: this requires the speaker to give clear instructions and the writer to listen carefully.

Improve learning and performance: critically evaluate own work

It is important that children have a reasonable sense of how well they are achieving the set goals and how they are approaching their work. In the plenary, they discuss their recordings with a partner, which provides a safe structure in which to assess their work.

Finding differences

Classroom technique: Heads or tails

Learning objectives

Maths
Find the difference between a pair of four-digit numbers

Speaking and listening
'Use precise language when learning'
Use precise language to explain ideas or give information

Personal skills
'Take pride in your work'
Improve learning and performance: take pride in work

Words and phrases
subtract, take away, difference, approximate, round, result, answer, check

Resources
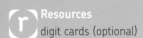
digit cards (optional)

Methods
Encourage children to choose any method with which they are comfortable and which is also
– appropriate (whether mental or written)
– reliable (produces the correct answer)
– efficient (can be carried out reasonably quickly).

Rounding and approximating
5407 4182
Round to 5400 and 4200, with a difference of 1200. Ask whether children think it is worth finding the exact difference. See if they will move on to trying another arrangement of the digits.

Heads or tails
Each pair must agree a clear and precise explanation and be prepared to give it to the class.

Introduction

Display these six digits and ask individuals to make two three-digit numbers using all the digits and to find the difference between them.

| 1 | 3 | 6 | 8 | 9 | 5 |

Collect in pairs of numbers children have found which have a difference close to 300.

Record the pairs of numbers and the differences.

Numbers		Difference
695	381	314
891	563	328

Briefly talk about the methods children have used for finding the differences and take the opportunity to do any teaching or revision of methods.

How did you find the difference between those numbers? Why did you choose that method?

How can we decide which difference is closest to 300?

Pairs

Write up these eight digits:

| 1 | 2 | 4 | 4 | 5 | 0 | 7 | 8 |

Working in pairs, children use these digits to make two four-digit numbers which have a difference as close to 1000 as possible. They record the pairs of numbers and the differences.

Stop the class after a while for a mini-plenary. Discuss how rounding the numbers will help them find an approximate difference.

When pairs believe they have found the closest possible difference, they agree how to describe the steps they took to arrive at the two numbers and how they found the difference.

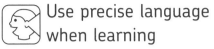

How do you find the approximate difference between two four-digit numbers? And how does that help with this problem?

When you are aiming for a difference of 1000, which part of the number do you consider first?

Those two numbers have a difference of 954. Can you adjust them to get a difference even closer to 1000?

Support: Use six digits, aiming for a target difference of 100. Use digit cards to rearrange the digits.

Extend: Use eight digits to make numbers with two decimal places, aiming for a target difference of 10.

Plenary

Using 'Heads or tails' (p12), ask pairs what they did to find two numbers with a difference close to 1000.

Encourage children to talk to their partners about how well they think they have worked and what they have achieved.

What did you look for when you were rearranging the digits?

How did this pair calculate the difference between two numbers?

Tell your partner one thing you have achieved in this lesson that you are proud about.

Assessment for learning

Can the children

Find the approximate difference between two four-digit numbers?

Explain why they tried particular combinations?

Name one thing they have achieved in the lesson that they are proud of?

If not

Ask children to round two three-digit numbers to the nearest 100 in case they don't understand how rounding numbers up or down enables them to find an approximate difference. Discuss how this is helpful in finding the approximate difference – for example, 476 – 241 becomes 500 – 200.

Ask children leading questions to help them think through their reasoning.

Make a habit of telling children what you think they have to be proud of and also tell them where you think they can improve.

Adding decimal fractions
Classroom technique: Peer tutoring

Learning objectives

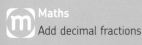

Maths
Add decimal fractions

Speaking and listening
'Reach an understanding with your partner'
Reach a common understanding with a partner

Personal skills
'Think about what you have learned'
Improve learning and performance: reflect on learning

Words and phrases
decimal, increase, decrease, total, sum, pattern, possibilities, solution, check, convince

Resources
display copy of RS12
RS13
for each pair:
copy of RS12

Peer tutoring
Put together pairs so that children who are weak at adding decimals are supported by children who are confident in this area.

Methods used
– jotting
– sketched a number line
– used my head

The problem-solving process
If children are not confident problem solvers, go through the following steps with them:
1. Read the problem to yourself.
2. Read the problem to your partner.
3. Discuss with your partner what you have to find out.
Take class feedback on what the children think the problem requires of them.

Mini-plenary
If children are to find all possibilities, they need to recognise that if one number increases by 1, another number must decrease by 1 – see RS13.

Introduction

Display the first grid from RS12.

Children find the sum of the four corners, then discuss with their neighbour the addition strategy they used.

Take feedback on the methods used and record them on the board.

Display the second grid on RS12 and discuss whether children would use the same methods to add the corner numbers here.

(m) *Which method do you prefer to use? Why?*

(m) *When might you do the calculations in your head? And when not?*

(s) *Tell your partner how you added those four numbers. What did you do first? And then?*

Pairs

Pairs tackle the problem posed on the second grid on RS12.

After about 10 minutes, stop the class for a mini-plenary. Take feedback from the children, show the numbers used on blank grids and record the related calculations.

	1.6	1.7	
	4.6	4.7	

Original corner numbers: 1.5 + 1.8 + 4.5 + 4.8 (total 12.6)

New numbers: 1.6 + 1.7 + 4.6 + 4.7 (total also 12.6)

Ask for ideas about any patterns the children notice.

Pairs continue with the problem, aiming to find all the possibilities. They also agree an explanation about how they know they have found them all.

(s) *Tell your partner why you are sure that you have found all the different possibilities.*

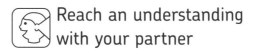 Reach an understanding
with your partner

 How do you think you are doing at adding decimal numbers? Are you getting quicker? Or more accurate?

Support: Aim to have all children working on the second grid. If necessary, children can work with the first grid, where finding the totals does not involve any 'carrying'.

Extend: Children find a way to extend the problem.

Plenary

Invite a few children to explain why they think they have found all the possible sets of numbers.

Then ask for ideas, and scribe them, about what children have been learning in the session – include personal and speaking/listening skills as well as mathematical ones.

> **What we have been learning**
>
> • adding decimals   • to work carefully
>
> • to check our work   • to help each other

At the end, children reflect in silence on what they as individuals have learned in the lesson.

 Is there anything you would like to add to Sarah's explanation? Or do you have any questions about it?

 Tell your partner if there is anything you particularly noticed them learning, or practising, or trying hard at, today.

Assessment for learning

Can the children

 Choose and use an efficient method of adding the decimal numbers?

 Be equally prepared, as partners, to offer an explanation of their work?

 Identify something they have been learning in the lesson?

If not

 Make more use of 'Peer tutoring' (p8) to teach these skills. Make decimals the focus of teaching for a while, emphasising the ways in which they operate like any other number in our base-ten system.

 Make use of techniques such as 'One between two' (p8) and 'Heads or tails' (p12) which focus pairs on equal participation.

 Give children feedback yourself, based on your observations in the lesson.

Calculating with money

Classroom technique: Talking partners

Learning objectives

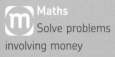 **Maths**
Solve problems involving money

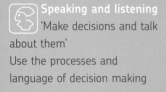 **Speaking and listening**
'Make decisions and talk about them'
Use the processes and language of decision making

Personal skills
'Plan and manage a group task'
Organise work: plan and manage a group task

W Words and phrases
price, cost, total, value, approximate, round, maximum, amount, decision, budget

r Resources
display calculator for each group:
copy of RS14
calculators

Checking the approximations
You could demonstrate with a calculator that adding the exact amounts gives a result close to the approximate total – so exact working is not necessary.

Arranging groups
Put together groups of mixed ability so that less confident children are supported by their peers.

Issues to discuss
– What are the different things children need to do?
– How will they share these out?
– How will they support each other and check they are each doing what was agreed?

Using rounding
Children will find rounding useful again to keep track of approximate costs. They will then need to work out exact amounts to check they do not exceed the agreed total for a toolkit.

Introduction

Display these amounts and ask individuals to find three amounts that total approximately £20:

£0.75	£2.45	£5.29	£6.85	£10.50	£14.00

Take suggestions and record them on the board. Emphasise that rounding is an essential tool when finding approximate totals. If each amount is treated as a whole number of pounds, finding an approximate total of £20 is relatively straightforward.

> £0.75 → £1
> £5.29 → £5
> £14.00 → £14
> Approximate total: £20

(m) *How did you decide which three amounts to try?*

(m) *How does rounding the amounts help?*

(m) *How close does 'approximate' mean?*

Groups of three

Groups of children share a copy of RS14.

Give children a few minutes to read and absorb the problem, then discuss with the class some of the issues they will need to deal with when sharing the work.

Children then answer the questions on RS14.

When the groups have agreed the contents of their toolkits, they check their work and find a way to present it clearly on paper.

(m) *How will you ensure you don't go over budget?*

(speaking) *How are you solving this problem? Do you like this way of doing it?*

(personal) *What are the different strengths of the people in your group? How can you make good use of those?*

(personal) *In your presentation, do you need to show what the toolkits cost in total? How can you do that?*

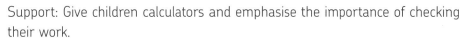
Support: Give children calculators and emphasise the importance of checking their work.

Extend: Say that you expect the costs of the kits to come close to the maximum allowed.

Plenary

Display each group's work around the classroom for children to look at the completed work.

Then have a class debrief of the experience of organising their own groups.

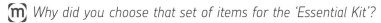 *Why did you choose that set of items for the 'Essential Kit'?*

 *If you were doing this again, would you organise your group the same or differently? Why?*

Assessment for learning

Can the children	If not
Use rounding to help them find approximate totals?	Practise this with the class, always comparing the results with an exact total on a calculator, to convince the children how effective the rounding technique is.
Explain their decisions about what to put in a toolkit?	Invite children who are confident do this to explain their decisions to the class.
Fulfil the brief as set out on RS14?	Work with the group to discuss and agree each child's role: Child A will choose an item, Child B will add it to the total, and Child C will say how much is left to spend.

Inverse operations

Classroom technique: One between two

Learning objectives

Maths
Use inverse operations to solve number puzzles

Speaking and listening
'Listen well'
Listen with sustained concentration

Personal skills
'Evaluate your own work'
Improve learning and performance: critically evaluate own work

Words and phrases
inverse, opposite, symbol, substitute, puzzle, problem, solution, check

Resources
for each pair:
copy of RS15
calculator (optional)

Inverses
Emphasise that each stage of the calculation was 'undone' by doing the inverse operation (encourage children to use this word, too). So if your last step was to add 9, their first step is to subtract it.

One between two
Child A devises the first puzzle and tells Child B what to record on the sheet. They then swap roles.

Calculators
If children want to challenge each other with big numbers, make calculators available – this activity is about understanding relationships between operations, rather than skill at calculating.

Using numbers and symbols
Encourage children to form equations using numbers and symbols, as you did in the introduction.

Introduction

The class will solve the following number puzzle: "I'm thinking of a number. First I double it. Then I add 9. The answer is 45. What is the mystery number?"

Ask how children worked it out and record an explanation on the board in numbers and symbols. Then check the solution.

> $(\square \times 2) + 9 = 45$ or $(m \times 2) + 9 = 45$
>
> Work backwards by taking off the 9:
>
> $45 - 9 = 36$
>
> Then halve the answer:
>
> $36 \div 2 = 18$
>
> Check 18 is correct:
>
> $(18 \times 2) + 9 = 45$

Revise the use of brackets if appropriate.

m *You suggested halving and taking away 9. Does it matter which you do first? Why?*

m *If I subtract 4 in my puzzle, what is the inverse operation you need to do?*

Pairs

Each pair works together to make up four number puzzles and record them on RS15, using 'One between two' (p8).

Pairs swap their sheet of four puzzles with another pair and solve the puzzles on the sheet they receive, again using 'One between two'.

m *How would you record 'half my number is 3'?*

Can you be sure you have written what your partner meant you to?

Are you setting puzzles that the other pair will find challenging?

How easy or hard do you find this? What bits do you find hard?

Support: Children practise the operations of addition and subtraction.

Extend: Children devise puzzles involving three steps.

Plenary

One or two children read out a puzzle they solved. Another child records the puzzle on the board and solves it. Children discuss this recording with their partner and think about other possible ways of tackling the problem.

(m) *What knowledge are you using to find the mystery numbers?*

(🗩) *Do you agree with what Sean said about his way of recording? Can you say why not?*

Assessment for learning

Can the children

(m) Choose inverse operations to solve puzzles which involve missing numbers?

(🗩) Repeat what their partners have just said?

(☺) Think about their work and say if they did well?

If not

(m) Check whether children can manage one-step puzzles, using all four operations. If so, build on this to working with simple two-step puzzles. If they do not understand the idea of inverse operations, work with a squares grid (for multiplication and division) or a number line (for addition and subtraction).

(🗩) Play games that require listening skills, such as 'Odd one out': one player gives a list of words or objects which have something in common and includes one that doesn't for the others to spot.

(☺) Do more work focusing on self-evaluation, showing that you value it.

Self and peer assessment

Lesson 9: Finding differences	I think	My partner thinks
(m) I can find the difference between two four-digit numbers.		
I can explain how to find the difference between two four-digit numbers.		
I am proud of the following work I did in this lesson: _____ _____		

Lesson 10: Adding decimal fractions	I think	My partner thinks
(m) I can use an efficient method for adding four decimal numbers.		
I am prepared to offer an explanation of our work to the class.		
I have learned the following in this lesson: _____ _____		

Name _____

Lesson 11: Calculating with money	I think	My partner thinks
(m) I can use rounding to help me find approximate totals.		
I can explain why we decided to put certain things in a toolkit.		
I found a way to work together with my group on a complicated task.		

Lesson 12: Inverse operations	I think	My partner thinks
(m) I can solve puzzles like the ones we did in today's lesson.		
I listen to what my partner tells me when we are doing 'One between two'.		
I think about my work and say if I did well.		

Self and peer assessment

Multiplication and division

Learning objectives

	Lessons			
	13	**14**	**15**	**16**
Ⓜ Maths objectives				
know by heart all multiplication facts up to 10 × 10	●			
use related facts and doubling and halving to multiply		●		
divide three-digit numbers			●	
understand multiplication				●
Ⓢ Speaking and listening skills				
explain and justify thinking	●			
share and discuss ideas and reach consensus		●		
listen with sustained concentration			●	
contribute to small-group discussion				●
Ⓟ Personal skills				
improve learning and performance: assess learning progress	●			
work with others: work cooperatively with others		●		
organise work: use different approaches to tackle a problem			●	
improve learning and performance: reflect on learning				●

About these lessons

Lesson 13: Multiplication facts

 Know by heart all multiplication facts up to 10 × 10

In this game, children choose two or three numbers to multiply together, in order to make numbers shown on a grid. This can help them memorise multiplication facts and, just as important, develop the sense of numbers as made up from smaller numbers, known as factors.

 Explain and justify thinking

Classroom technique: One between two

Pairs of children work as a team to play the game. One child makes the decisions and instructs their partner which moves to make. They must also give the reasons for their decisions.

 Improve learning and performance: assess learning progress

The game highlights which facts children do, and don't, know. In the plenary, a test reveals further information on what facts children are weak on, and they are asked to note these facts down and learn them.

Lesson 14: Doubling and halving strategy

 Use related facts and doubling and halving to multiply

In this activity, children think about whether a multiplication calculation can be made easier by doubling one number and halving the other. Using doubling and halving in this way to simplify a calculation is a valuable addition to children's repertoire of multiplication strategies.

 Share and discuss ideas and reach consensus

Classroom technique: One between two

Pairs of children discuss strategies for doing one of a series of multiplications. One child then instructs the other who has the pencil what to write in order to carry out the calculation, using a combination of mental working and jottings.

 Work with others: work cooperatively with others

Children share a task and discuss how to fulfil it. In addition, as they share just one pencil, they must cooperate over its use.

Lesson 15: Division of three-digit numbers

 Divide three-digit numbers

Children need to understand, be confident and competent with an efficient, informal, written method of division. In this activity, children practise their division skills.

 Listen with sustained concentration

Classroom technique: Ticket to explain

In the plenary, children earn the right to offer their own idea by re-explaining what the previous child has just said. This encourages them to listen carefully to what others are saying.

Organise work: use different approaches to tackle a problem

Children can approach the problem they are presented with in whatever way they choose. A few minutes into the lesson, a mini-plenary allows for a sharing of ideas, giving children the opportunity to change or adapt their approach to the problem.

Lesson 16: A multiplication problem

 Understand multiplication

Children explore the number of ways of combining different categories of objects – in this case, sandwiches and fillings – and learn how to use multiplication to solve the problem. This kind of problem gives children the foundations for later work in combinations and probability.

 Contribute to small-group discussion

Classroom technique: Think, pair, share

Children think about a problem alone, then discuss their ideas in pairs and groups of four. This structure gives them an opportunity for independent thought as well as the benefits of sharing ideas. It also allows less confident children time to prepare something to say in the group discussion.

Improve learning and performance: reflect on learning

In the plenary, specific questions encourage children to think about what they have gained from the lesson.

Multiplication facts
Classroom technique: One between two

U&A Interpreting data in tables
Make sure children understand that, on the grid, many numbers appear twice or more. So if their dice give them a product of, for example, 24, they have a choice of 4 squares to cross out.

All the possible products
Work with the class to establish all the products that can be made by multiplying two or three of the dice numbers together:

2 4 5
2 × 4 = 8
2 × 5 = 10
4 × 5 = 20
2 × 4 × 5 = 40

Speaking and listening
Ban pointing. Instead, encourage children to say things like: "We'll multiply the 4 and the 5 so we can cover the 20 in the fours column."

Introduction

Play a game against the class on a display copy of RS16 (show one grid only). The aim is to cross out a line of five squares.

Roll two 1–6 dice and one 0–9 dice and write up the numbers. Pairs choose two of the numbers, or all three, multiply them together and write their answer on a wipe board.

Scan the class's wipe boards to see what product the majority want and ask children to help you find that number and draw a cross on it in the class's colour. Make a point of discussing why that number was chosen. (Unless the product is a square number, it will appear at least twice – establish which of the number's appearances children want you to cover.)

Then you have your turn, using your own colour pen.

Continue like this until one team has a line of 5 squares crossed through in their colour.

1	2	3	4	5	6	7̶	8̶	9	10
2	4	6	8̶	10̶	12̶	14̶	16̶	18	20
3	6	9	12	15̶	18	21̶	24	27	30̶

ⓜ *I'm multiplying 4 and 6. How can I use the numbers along the top and down the side to help me find their product?*

ⓜ *Would 32 appear in a column for multiples of 12?*

🗣 *Why do you want to choose the 3 and the 4 to multiply?*

Pairs/Groups of four

Children play in teams of two against each other, using a copy of RS16. Child A rolls the dice; Child B decides what multiplication to do and tells Child A which square to cross through. If Child A disagrees with, or is uncertain of, a decision, they must ask for the reasoning behind it and discuss together what to do.

They then swap roles, playing two games in all.

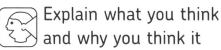
Tell your partner why you want to cross out the 35 in the times 7 row.

Tell me some numbers you can multiply easily. Tell me some you cannot multiply so easily.

Support: Organise mixed-ability pairs. Make sure children understand how to read the grid to find a product.

Extend: Remove the top row and first column. Alternatively, use RS17.

Plenary

Make sure the multiplication grids are hidden from view.

Announce two numbers. Pairs quickly multiply them together and write their answer on a wipe board to show you.

Any pairs who are slow or who give the wrong answer jot down the pair of numbers as a reminder that they need to learn their product.

m *What other two numbers have a product of 18?*

Which tables do you think you need to practise?

Taking responsibility
By Year 6, children should be able to see for themselves the importance of knowing the multiplication facts up to 10 × 10 and take on the task of identifying and learning 'problem' ones.

Assessment for learning

Can the children

m Say which dice numbers they could use to make 16, 24, 35, 64?

Talk about why they chose to multiply those particular numbers in the game?

Identify what multiplication facts they need to learn?

If not

m Work on producing factor trees.

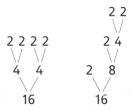

Ask children to listen to and repeat the reasons given by the other pair they are playing with.

Ask pairs of children to share responsibility for each of them learning a particular set of multiplication facts (for example, the 6 multiplication table). Encourage them to test each other, analyse which facts are problematic and find ways of remembering them.

Doubling and halving strategy

Classroom technique: One between two

Doubling and halving

This may not be the only strategy for a multiplication, but emphasise that it is worth checking whether this strategy might work, as it can make a calculation quick and easy. For example, 15×12 can be done easily by partitioning, whereas doubling and halving turns it into 30×6, which can be calculated even more easily.

Checking calculations

Allow children to check their calculations with a calculator. This means they can spot where they have made an error and try to trace it.

Formal recording

The sample recording in the worked example on RS19 shows the correct use of brackets. As this exercise is about mental strategies and jottings, tell the children they need not show brackets as long as their method is clear and easy for themselves, or you, to check.

Introduction

Display RS18. Children work in pairs to identify pairs of multiplications (one from each column) that have the same product and discuss which of the two calculations is easier to do.

Ask volunteers to identify the pairs and work as a class to carry out the multiplications, focusing on the strategy of doubling one number and halving the other.

Check the calculations with a calculator or let children do this on their own calculators.

> 17×14 looks hard, but doubling and halving turns it into a calculation that can be done fairly easily with partitioning:
>
> $17 \times 14 = 34 \times 7$
> $ = (30 \times 7) + (4 \times 7)$
> $ = 210 + 28$
> $ = 238$

(m) *Which is easier, 28×15 or 14×30? Why is it easy to multiply by a number with a zero in the units?*

(s) *Talk with your partner about whether doubling one number and halving the other will always give you the correct answer.*

Pairs

Each pair of children works with a copy of RS19 and one pencil. Child A picks a calculation to do. Both children discuss whether the doubling/halving strategy would work and whether there is another straightforward strategy that they could also use. Child A (who makes the ultimate decision on strategies) then instructs Child B (who has the pencil) what to write in the grid, following the style of the worked example.

Children swap roles for the next calculation.

(m) *Why do the multiplications where you can use doubling and halving all have at least one even number?*

(m) *Explain if doubling or halving could make this division any easier:*
126 ÷ 14

(☺) *Is it OK for you to disagree with your partner? And what will you write in the grid if you disagree?*

Support: Direct children to the first five problems, which are relatively easy. Make sure they have access to a multiplication grid.

Extend: Give children a calculator and ask them to explore whether there is a strategy that involves doubling and/or halving that works for division.

Plenary

Display a copy of RS19. Pairs contribute to a calculation they themselves did and instruct you how to add it to the chart. Before they describe what they did, pairs indicate, using 'thumbs up' or 'thumbs down', whether halving and doubling worked and whether they found a different 'easy' method.

Children then turn to their partners and tell them one way in which they have been cooperative and pleasant to work with in this lesson.

(m) *Is there anything special about those numbers where you can use halving and doubling?*

(m) *Could a division calculation be made easier by doubling or halving?*

(🗪) *Talk to your partner about the skills you need to be able to do these calculations in your head.*

Assessment for learning

Can the children

(m) Use the doubling/halving strategy to do calculations mentally?

(🗪) Tell you whether they agree or disagree about the strategies for solving a problem?

(☺) Accept if their partner chooses a strategy they do not agree with?

If not

(m) Work on multiplication table facts if children do not have a good enough knowledge of them. They also need to understand that multiplying by, for example, 30 is just as easy as multiplying by 3.

(🗪) Ask children to write an 'A' (agree) or 'D' (disagree) after each calculation to indicate whether or not they agree about it.

(☺) Point out that children will have the right to decide on the strategy next time and, meanwhile, urge them to accept their partner's right to make a decision when it is their turn.

Division of three-digit numbers

Classroom technique: Ticket to explain

Learning objectives

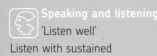

Maths
Divide three-digit numbers

Speaking and listening
'Listen well'
Listen with sustained concentration

Personal skills
'Try different ways to tackle a problem'
Organise work: use different approaches to tackle a problem

Words and phrases
factor, multiple, product, divisor, quotient, divisible, problem, strategy, method, solution, check

Resources
display copy of RS20
calculator
for each pair:
copy of RS21

Entering a problem
Children follow these steps:
1. Read the problem.
2. Read it to a partner.
3. Tell each other what you think you have to find out.

Using calculators
Tell children they need to put the unknown at the end when they use a calculator to do a calculation.

Sketching
Children can sketch the first few floors of a skyscraper to help them visualise the problem and find a way towards its solution.

Introduction

Display a copy of RS20. Working in pairs, children discuss the problem and write an appropriate number sentence, then find the answer.

Collect in suggestions as to the appropriate number sentence. If children offer the number sentence $15 \times \boxed{} = 375$, acknowledge that this is valid, but ask them what operation they did to find the answer. Establish that it was division (even if they used multiplication to find the answer) and reformulate the calculation as $375 \div 15 = \boxed{}$.

m *How do we know this is a division problem?*

Speaking and listening *What did Connor just say? Do you agree with him?*

Pairs

Each pair of children works with a copy of RS21, starting with Question 2. Children record the number sentence for each number of windows and floors they try and the calculations (unless they take a short cut using their knowledge of divisibility).

$162 \div 7$	$162 \div 8$
162	8 won't work because you can't halve 162 twice. So the skyscraper can't have 8 windows on a floor.
$-\ 70\ \ (10 \times 7)$	
92	
$-\ 70\ \ (10 \times 7)$	
22	
$-\ 21\ \ (3 \times 7)$	
1	

$162 \div 7 = 23$ r 1, so the skyscraper can't have 7 windows on a floor.

After 10 minutes, stop the class for a mini-plenary and discuss with the children what the problem requires them to do. Establish that they need to try various numbers of windows (or floors) to see which numbers divide into 162 with no remainder. Briefly discuss methods for doing this.

Pairs then return to their problem.

(m) *The number divides by 6, so what else will it definitely divide by?*

(m) *How can you tell 162 divides by 9, without even doing the calculation?*

(☺) *Is there another way you could tackle this problem?*

Support: Children work in pairs on the first question on RS21.

Extend: Children try to find all the possible solutions before going on to tackle Question 3. Consider asking these children to act as peer tutors to children who are struggling.

Dividing 162
Numbers that divide:
6 and 27, 9 and 18
Numbers that don't divide:
7 8 10 11 12
13 14 15 16 17

Ticket to explain
Children earn the right to offer their own idea by re-explaining what the previous child has just said.

Plenary

Children help you use a calculator to test which numbers of floors or windows work for a total number of 162 windows. Record those numbers that work and those that don't. Ask for explanations of what you can deduce from your results.

Then ask children for the methods they used to work out their division calculations, using 'Ticket to explain' (p11).

Record the children's explanations.

(m) *If we know that 162 is divisible by 9, how does that help us answer the problem?*

(m) *What has this problem got to do with finding factor pairs?*

(☺) *Can you explain Fatima's idea about why there is no point looking for factors larger than 27?*

Assessment for learning

Can the children

(m) Choose and use an accurate method for dividing a three-digit number by another number?

(☺) Repeat an idea or explanation offered by another child?

(☺) Adapt their problem-solving strategy if appropriate?

If not

(m) Revise using multiples of the divisor as a method for division, starting with two-digit numbers.

(☺) Do listen-and-repeat activities in pairs, using a variety of topics such as sport, holidays, pets or favourite books.

(☺) Try to establish if children feel insecure with the division aspect or the problem aspect of the activity. If this is the case, provide work at a comfortable level and explore different methods for reaching a solution.

A multiplication problem

Classroom technique: Think, pair, share

Learning objectives

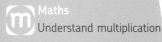 **Maths**
Understand multiplication

Speaking and listening
'Join in a discussion with a small group'
Contribute to small-group discussion

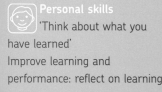 **Personal skills**
'Think about what you have learned'
Improve learning and performance: reflect on learning

 **Words and phrases**
product, represent, combination, groups of, method, table, diagram, systematic

Resources
display copy of RS22

Think, pair, share
Use this technique (p10). Children think about the problem alone for one minute before discussing it in pairs for a further minute. Pairs then join up in groups of four to discuss briefly and ensure that each member of the group gets to speak and is listened to respectfully.

Working unsystematically
Children may approach the problem and record possible solutions in a fairly unsystematic way. Give them time to explore this approach and discover its pitfalls.

Think, pair, share
Use this technique again.

Introduction

Display a copy of RS22.

Children briefly talk about the problem.

As a class, discuss ways of tackling the problem. Pairs of children then work on it, recording their thinking in whatever way they choose.

Stop the children after 10 minutes and collect in suggestions from pairs who have had some success with the problem. Record the children's work on the board. If no one has come up with a matrix, introduce this way of recording the combinations, using agreed abbreviations for the sandwich types.

	white	brown	French
cheese	ch/wh	ch/br	ch/Fr
peanut butter	pbutter/wh	pbutter/br	pbutter/Fr
chicken	chick/wh	chick/br	chick/Fr
tuna	t/wh	t/br	t/Fr

If we want tuna filling, what types of bread could we have? How many different types of tuna sandwich are possible?

Can anyone suggest a more systematic way of working?

What are the pros and cons of the different methods on the board?

Pairs

Pairs of children explore and record what happens if Lucca adds bagels to the list of breads and roast beef to the list of fillings.

Before the plenary, pairs compare their results. Present the class with the question: "What has this problem got to do with multiplication?"

You are using a table. What do you need to have along the top here?

Tell your group why you think multiplication can help with this problem.

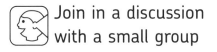

Have you got all the possible combinations? How can you be sure?

Did you learn anything from what this other pair said?

Support: Children solve the problem for 2 breads and 3 fillings.

Extend: Children solve the problem for 7 breads and 9 fillings or for 12 breads and 17 fillings. Ask for a general explanation that works for any number of breads and fillings.

Plenary

Share and discuss the children's findings. If some children are still thinking about the problem additively, help them move towards thinking multiplicatively.

> There are 12 combinations:
> 4 white bread + 4 brown bread + 4 French bread
> 4 fillings × 3 breads = 12 combinations

Pairs of children work out a few sandwich combinations involving larger numbers.

At the end, children 'tell their partner' (p10) two things they have been learning about in this lesson.

Does it matter whether you use addition or multiplication to solve this problem? Which is quicker?

Is there anything you can do now that you couldn't do at the start of the lesson?

Would you be able to use multiplication to solve other problems like this?

Assessment for learning

Can the children

Solve the problem for 3 breads and 5 fillings or for 10 breads and 17 fillings?

Explain how they solved the problem?

Identify two things they have been learning about in the lesson?

If not

Help children model the problem and solution with objects, working and recording systematically, then let them do the same for a simpler version of the problem.

Use the child's recording to form your own explanation of their method and have them repeat in their own words what you say.

Ask other children to offer suggestions, which the child can agree with or not.

Name _____

Self and peer assessment

Lesson 13: Multiplication facts	I think	My partner thinks
(m) I know all the multiplication facts in these tables: 2× 3× 4× 5× 6× 7× 8× 9× 10×		
I can explain why I chose a particular product when playing the grid game.		
I know what multiplication facts I need to learn next.		

Lesson 14: Doubling and halving strategy	I think	My partner thinks
(m) I can see when doubling and halving can help with a multiplication calculation.		
I can say whether I agree or disagree with my partner about how to solve a problem.		
I am happy for my partner to choose a method I don't like.		

Name _____

Lesson 15: Division of three-digit numbers	I think	My partner thinks
(m) I can divide a three-digit number by another number.		
I can repeat an idea that someone else has just said.		
I am prepared to change my problem-solving strategy if it isn't working.		

Lesson 16: A multiplication problem	I think	My partner thinks
(m) I can solve the sandwich problem.		
I can explain how I solved the sandwich problem.		
I can name two things I have been learning about in this lesson.		

Self and peer assessment

Handling data

Learning objectives

	Lessons			
	17	**18**	**19**	**20**
ⓜ Maths objectives				
extract and interpret data in tables and graphs	●			
draw and interpret a graph		●		
interpret data in tables and graphs			●	
organise data on a Venn diagram				●
Ⓢ Speaking and listening skills				
share and discuss ideas and reach consensus	●			
listen with sustained concentration		●		
contribute to small-group discussion			●	
explain and justify thinking				●
Ⓟ Personal skills				
organise work: plan ways to solve a problem	●			
work with others: discuss and agree ways of working		●		
work with others: work cooperatively with others			●	
improve learning and performance: develop confidence in own judgements				●

About these lessons

Lesson 17: Reading data in charts

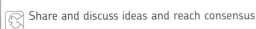 **Extract and interpret data in tables and graphs**

The best way for children to understand graphs and tables is to use the data to solve a 'real-life' problem. Here, children use both a table and a conversion graph to help them answer a series of questions.

Share and discuss ideas and reach consensus

Classroom technique: Think, pair, share

At several points in the lessons, children think about a problem as individuals, then share their thoughts with a partner as well as in groups of four, aiming to reach agreement on the answer.

 Organise work: plan ways to solve a problem

Children need to find their way through a problem, but are supported by working both with a partner and in a larger group.

Lesson 18: Drawing a line graph

 Draw and interpret a graph

Children need experience of the process of constructing a line graph and making decisions about the scales involved. They also need to practise interrogating graphs to extract information. Here, children work together to draw a line graph that tells a story, then write statements about it.

Listen with sustained concentration

Classroom technique: Ticket to explain

In the introduction and plenary, children 'earn' the right to speak by repeating the explanation given by the previous child. Inform the class that you will be using this technique as an incentive for them to listen carefully to their peers.

Work with others: discuss and agree ways of working

As pairs of children share the task of constructing the graph, they must find ways of sharing the work equally.

Lesson 19: Interpreting graphs

 Interpret data in tables and graphs

In this activity, children look at data represented in tables and graphs and work out what, if anything, is wrong with the way it is displayed.

Contribute to small-group discussion

Classroom technique: Think, pair, share

Children consider a problem individually. They then explain their ideas to a partner. After pairs have discussed the issue, they join another pair for a wider airing of views.

 Work with others: work cooperatively with others

Children share a task with other children, being aware that any of them may be called on to speak in the plenary. This gives all of them an incentive to participate in the task and cooperate with the other group members.

Lesson 20: A Venn diagram

Organise data on a Venn diagram

Children sort numbers onto a Venn diagram showing two intersecting sets. This allows children to explore relationships between numbers, such as the fact that any number which is a multiple of 4 and of 3 is also a multiple of 6 or that 2 is the only even prime number.

Explain and justify thinking

Classroom technique: Heads or tails

Pairs of children formulate statements about their discoveries, knowing that either of them may be called on to give their explanations in the plenary – so both take equal responsibility for understanding these explanations.

 Improve learning and performance: develop confidence in own judgements

Choosing their own properties to sort by allows children to work at a level where they feel comfortable. This helps them develop confidence in the results of their sorting.

Reading data in charts

Classroom technique: Think, pair, share

Learning objectives

m Maths
Extract and interpret data in tables and graphs

Speaking and listening
'Share ideas and reach agreement'
Share and discuss ideas and reach consensus

Personal skills
'Plan ways to solve a problem'
Organise work: plan ways to solve a problem

W Words and phrases
conversion graph, mileage chart, horizontal, vertical, axis, axes, approximate, equivalent, kilometre, mile

r Resources
display copy of RS23
display copy of RS24
for each pair:
copy of RS23, RS24 and RS25
rulers
calculators

Think, pair, share
Children think for a minute about the questions, then discuss their ideas with a partner. Finally, they join another pair for further sharing.

Converting distances longer than 50 miles
Children might break down the journey into chunks and work out the equivalents for each bit or work out the kilometre equivalent for half the journey and double the result.

Mini-plenary on the problem-solving process
Children follow these steps:
1. Read the problem to yourself.
2. Read the problem to your partner.
3. Discuss with your partner what you have to find out.
It is important then to discuss as a class how the problem needs to be addressed.

Introduction

Display RS23 (or give children their own copies). Working in pairs, children look at the chart and discuss the questions.

Clarify with the class how to use the chart.

Display RS24 (or give children their own copies) and explain that 5 miles is equivalent to approximately 8 kilometres.

Children look at the chart and discuss the questions.

Establish with the class how to use the graph and how to use a ruler to ensure they follow a line correctly. Agree to round values to the nearest mile/kilometre.

Distance between British towns and cities in miles											
Aberdeen											
468	Aberystwyth										
431	124	Birmingham									
606	285	171	Brighton								
514	128	90	169	Bristol							
532	110	109	201	44	Cardiff						
125	335	298	474	381	399	Edinburgh					
587	195	164	175	84	111	454	Exeter				
147	333	296	472	380	397	48	453	Glasgow			
106	494	458	633	541	558	157	614	173	Inverness		
387	199	89	216	186	204	259	260	279	440	Lincoln	
358	110	102	277	185	202	224	258	222	384	140	Liverpool

m *Explain how we can use this chart to find the distance between any two towns.*

⟨⟩ *With your partner, talk about how to convert 24 miles to kilometres. What can you do about the fact that the answer isn't an exact number of kilometres?*

Pairs/Groups of four

Pairs of children work with a copy each of RS23, RS24 and RS25, discussing the charts and questions. They then join up with another pair and agree the home town of the courier company. Each child in the group chooses one destination for the van to travel to. The group record the home town of the courier company and all four destinations.

Children then work in pairs again to discuss and solve the problems on RS25.

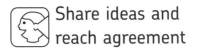 Share ideas and reach agreement

Sharing ideas and checking work
This is a variation of 'Think, pair, share' (p10).

When pairs have solved the problem, they get back into a group to share and check their results, discussing and correcting any discrepancies.

Do you both agree that the journey is 480 km? Have you checked that you have answered the question?

What is a useful thing to do first when solving the problem?

Support: Enlarge the bottom left part of RS24 and work with this group to read and interpret the graph.

Extend: Children devise a ready reckoner for converting miles to kilometres.

Plenary

Sketch a table showing some distances in both miles and kilometres.

mile	km
10	16
20	32
25	40

The answer
Multiplying any number of miles by 1.6 gives an approximate number of kilometres; dividing any number of miles by 0.6 gives an approximate number of kilometres. Although it is useful for children to know this, don't expect them to reach a clear understanding of this in the time available.

Make calculators available. Use 'Think, pair, share' again to discuss the relationship between the numbers in each column. Ask whether there is a quick way to convert any number of miles to kilometres without using the graph.

Assessment for learning

Can the children

Read and round an intermediate value on the line graph?

Tell you something their partner has just said, and whether they agree with it?

Identify what they will do first to tackle the problem?

If not

Focus on the section from 0 to 10 miles, enlarged. Work together to position 5 km and 5 miles on the axes, find equivalent values for these and round them.

Play some maths games with pairs working together, discussing strategies and decisions. The simpler context of the game allows you all to focus on collaboration.

Join confident pairs with less confident ones, using 'Think, pair, share' (p10) at various stages of the problem solving.

Drawing a line graph
Classroom technique: Ticket to explain

Learning objectives

m Maths
Draw and interpret
a line graph

Speaking and listening
'Listen well'
Listen with sustained
concentration

Personal skills
'Discuss and agree how
to work'
Work with others: discuss and
agree ways of working

W Words and phrases
horizontal, vertical,
axis, axes, distance, time,
graph, scale, interpret, data,
information

r Resources
display copy of RS26
display copy of RS27
for each pair:
copy of RS27
1 cm squared paper
or graph paper

Introduction
Display RS26 and read through the story.

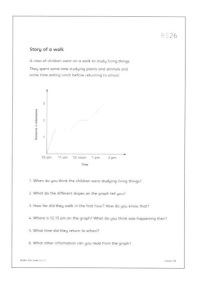

Class discussion
Use 'Ticket to explain' (p10).
A child must repeat the
explanation that has just
been given by the previous
child before answering their
own question.

Pairs of children discuss the questions. Then open out
the discussion to the class.

Tell me one fact you can work out from the graph.

Tell us what Rick just said about the first question.

Pairs

Pairs of children share a copy of RS27 and some
squared or graph paper. They agree on the time of day
that belongs in each empty space on the statements
describing the walk. They then use the information to
draw a distance/time graph of the long hike.

**What scale to use
for the axes?**
On the vertical axis, one
centimetre per kilometre
would work. On the horizontal
axis, two centimetres for each
hour is appropriate.

Pairs devise further statements (on a separate sheet of
paper) based on the information shown in the graph.

*How far did the children walk altogether? Why does
that information help you with drawing the axes?*

How will you share the work of drawing the graph?

Support: Work with this group. Show children how
to draw the axes and do the scales.

Extend: Pairs write true and false statements.

Plenary

Display RS27 and work with the class to sketch the graph of the long hike on
the board. Read out some of the children's statements based on the graph.
(Include some of the false statements produced by the extension group or
change some of the values in the statements other children have written.)

Pairs discuss whether each statement is true or false. Individuals then tell
you their answer, justifying their decision.

(m) *Now we have the axes drawn, tell me how to draw in the first part of
the walk.*

(m) *How do you know that statement is false?*

If you agree with what I have just said, show me a 'thumbs up'.

Devil's advocate
Discussing true and false
statements – a variation of
'Devil's advocate' (p11) – is a
useful device for encouraging
children to think and talk
mathematically.

Ticket to explain
Use this technique (p11).

Assessment for learning

Can the children

(m) Use the information given on RS27 to draw
a distance/time graph?

Repeat to you or their partner what someone
has just told the class?

Agree with their partner how to share the work?

If not

(m) Look at some more graphs that 'tell a story' with
unlabelled axes. With the children, agree labels for
the axes, then tell the stories that the graph shows.
Focus on precise interpretation.

Have brief sessions, on a daily basis, where
children repeat to a partner statements or
explanations given by you or another child –
these can be on any topic, not just maths.

Use 'One between two' (p8), which is a more
structured technique for sharing work.

Interpreting graphs
Classroom technique: Think, pair, share

Learning objectives

Maths
Interpret data in tables and graphs

Speaking and listening
'Join in a discussion with a small group'
Contribute to small-group discussion

Personal skills
'Work cooperatively with others'
Work with others: work cooperatively with others

Words and phrases
data, table, graph, interpret, explain, per cent, total, likely, unlikely, pie chart

Resources
display copy of RS28
for each pair:
copy of RS28

Think, pair, share
The procedure
– individual considers graphs
– pair shares thoughts
– two pairs join together and share and compare ideas

Reporting back
Remind children that, in the plenary, you may ask any member of the group for reasons for their chosen answers, so they should all be prepared.

Introduction

Have a brief discussion about school dinners and establish some data based on the class's experiences.

m *If ten of you eat school dinners, is that more or less than 50% of the class?*

m *How many school dinners will this class eat over the course of a term?*

Pairs/Groups of four

Pairs share a copy of RS28.

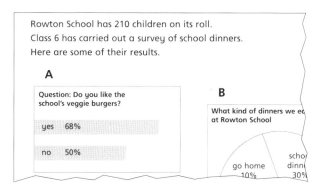

Rowton School has 210 children on its roll.
Class 6 has carried out a survey of school dinners.
Here are some of their results.

A

Question: Do you like the school's veggie burgers?

yes 68%

no 50%

B

What kind of dinners we ea at Rowton School

go home 10%   scho dinn 30%

Individuals look at all the data and think about it on their own. They then discuss with their partner what, if anything, is wrong with each graph or table*, aiming to reach agreement where possible.

Pairs join another pair and compare their ideas, making notes about what is said. They then go back to their pairs and write a brief statement for each graph, outlining their thoughts about it.

m *What is that chart telling us? And does that make sense?*

Explain to Marie how you think that pie chart should look.

Check what you have written and compare it with what the other pair has written.

Support: Work with this group.

Extend: Children correct or redraw the tables and graphs so they are coherent and accurate.

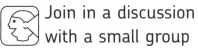
Plenary

1, 2, 3, 4
Using this technique (p12), roll a dice to determine which member of a group of four answers your question. Roll the dice again if you throw 5 or 6.

Choose a group of four. Display one of the charts and ask one member of the group to explain what is wrong with it.

Ask the class if they have any ideas on how to correct the chart or anything to add to the explanation given.

Repeat with other groups and other questions.

m *How is a pie chart meant to work?*

m *Did you think of any reason why there might be 163 school dinners on that day?*

---

* Points to make about the graphs and tables

A The two values for 'yes' and 'no' should add up to 100% (or less, if some children didn't answer).

B The three percentages add up to 100%, as they should do, but the whole pie is not divided up accurately – the 10% slice looks more like 20%.

C There is nothing really wrong with this, but it is unlikely that 163 children ate school dinner on Thursday in week 2, when normal figures are 60 to 70.

D The horizontal axis has no figures, so we cannot read from the graph how many children in each class ate dinners on that day. However, as normal numbers are around 60, some children may be able to work out that the scale should be marked in 2s and that, for example, 11 Year 6 children ate dinner that day.

Assessment for learning

Can the children

m Read data from the tables and graphs shown?

Take part in the discussion when pairs join up?

Share in the work of writing explanations?

If not

m Establish which charts children have problems with and do further work on these.

Consider introducing a 'talking stick' (p10), a decorated stick which confers on the holder the right – and responsibility – to speak and be listened to.

Use the technique 'One between two' (p8) which requires children to share one pencil, making cooperation a necessity.

A Venn diagram

Classroom technique: Heads or tails

Learning objectives

m Maths
Organise data on
a Venn diagram

Speaking and listening
'Explain what you think
and why you think it'
Explain and justify thinking

Personal skills
'Develop confidence
about what you think
and decide'
Improve learning and
performance: develop
confidence in own judgements

W Words and phrases
multiple, property,
record, data, Venn diagram,
label, explain, generalise,
predict, check

r Resources
multiplication grids
and 100-grids (optional)

Tell your partner
Children discuss with their
partner where the number
belongs each time. Ask for
suggestions and check
whether the other children
agree or disagree.

'Whats' and 'Whys'
What?
The square numbers go: odd,
even, odd, even.
The numbers outside the sets
are all odd.
Why?
The numbers you are squaring
go odd, even, odd, even, too.
If they were even, they would be
in the even set.

Heads or tails
Remind children that you will be
using this technique (p12) in the
plenary. So both members of the
pair should be prepared to talk
about what they have found out
from their sorting. Emphasise
that you want explanations
about 'why', not just descriptions
of 'what'.

U&A Testing statements
Encourage children to test their
statements by looking for
counter-examples: "Can we find
a number that is a multiple of
4 and of 3 but isn't a multiple
of 6?"

Introduction

Sketch a Venn diagram with two intersecting sets and brainstorm properties of numbers that the children could sort numbers by. Scribe these (a single property at a time) and add any that you want children to focus on.

odd	even	prime	square
multiple of 10		multiple of 3	

Choose two of these properties and label the two sets accordingly. Invite suggestions for numbers and put these onto the diagram with the help of the children.

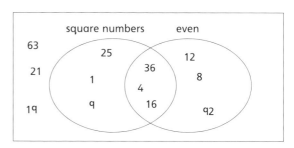

When several numbers have been sorted, children give explanations and comments. Scribe these to act as a model for children's paired work, distinguishing between 'whats' and 'whys'.

What can you say about the numbers in this section of the diagram?

What kind of numbers belong outside both circles?

If we sorted all the numbers to 100, which section would have most numbers? Why?

What do the numbers in this section have in common?

Pairs

Pairs draw their own Venn diagram and choose their own properties to sort by, preferably properties that might produce interesting results.

They then look for numbers to go in each part of the diagram. After sorting about 20 numbers, pairs discuss and record an explanation of their results.

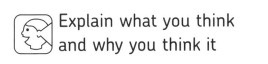

What kind of numbers go in this part of your diagram?

Tell me why you are sure about that.

How can you check that this is true?

Support: Make sure children choose properties they are certain of, such as 'even' and 'multiple of 3', and provide a multiplication grid and a 100-grid as a check.

Extend: Children label one of their sets 'prime numbers' and discuss what their sorting shows up about this set.

Plenary

Heads or tails
Toss a coin to decide which of the pair presents their results.

Ideas board
If children have done work they would like to share but have not had an opportunity to talk about in the plenary, let them display it on an 'ideas board' (p13) for other children to look at in the next few days.

Ask pairs to present their discoveries, using 'Heads or tails' (p12).

For each pair, sketch their sets on the board and add a few of the numbers they worked with, so the class can understand the presentation.

Can you say why there were no numbers in this part of your diagram?

You found only one of the numbers in 'prime numbers' also belonged in 'even'. Why is that?

Assessment for learning

Can the children

(m) Find the correct place on their diagrams to write their numbers?

Reach a conclusion about the results of their sorting?

Speak confidently when telling you or the class about their conclusion?

If not

(m) Go back to first principles if children have problems understanding about the significance of the intersection. Sort number cards into two separated hoops and talk about what to do with the cards that belong in both hoops.

Ask children to look at each section of the diagram and describe the numbers there. Make a statement yourself for them to agree or disagree with.

Help children formulate some statements about their sorting and check each one by looking for counter-examples: "Let's see if we can find any numbers that go there and aren't odd. No? Then your statement must be correct."

Name _____

Self and peer assessment

Lesson 17: Reading data in charts	I think	My partner thinks
(m) I can work out the value of any point on the line graph.		
I can repeat something my partner has just said and say whether or not I agree with it.		
I can decide what I will do first when starting on a problem.		

Lesson 18: Drawing a line graph	I think	My partner thinks
(m) I can draw a distance/time graph.		
I can remember and repeat what another pupil has just told the class.		
I try to agree with my partner how we will share the work.		

Name _____

Lesson 19: Interpreting graphs	I think	My partner thinks
(m) I can read data from the tables and graphs we looked at.		
I join in the discussion in our group of four.		
I share with my partner the work of writing explanations.		

Lesson 20: A Venn diagram	I think	My partner thinks
(m) I can find the correct place on the Venn diagram to write each number.		
I can say something that we have discovered from our sorting.		
I know which bits of my work I feel confident about.		

Self and peer assessment

Measures

Learning objectives

	Lessons			
	21	**22**	**23**	**24**
ⓜ Maths objectives				
calculate the area and perimeter of compound shapes	●			
read a measuring scale		●		
work with a scale plan			●	
solve problems involving timetables				●
🕲 Speaking and listening skills				
share and discuss ideas and reach consensus	●			
explain and justify thinking		●		
discuss progress of work			●	
reach a common understanding with a partner				●
☺ Personal skills				
work with others: work cooperatively with others	●			
improve learning and performance: assess learning progress		●		
organise work: work on a task with several aspects			●	
organise work: plan ways to solve a problem				●

About these lessons

Lesson 21: Area and perimeter

 Calculate the area and perimeter of compound shapes

Children often become confused when asked to find both the area and perimeter of a shape. In this activity, children draw compound shapes, then find their area and perimeter. Finally, they discuss why, when two shapes are combined, you can add the areas of each of them to find the area of the compound shape, but you cannot do the same with their perimeters.

 Share and discuss ideas and reach consensus

Classroom technique: Talking partners

In this activity, children work together informally, sharing the working and making decisions together.

 Work with others: work cooperatively with others

Children share a task with a partner, which cannot be completed unless they both cooperate. Pairs then swap work with another pair and work on a task set by the other pair – requiring further cooperation.

Lesson 22: Reading scales

 Read a measuring scale

In this activity, children read a scale showing masses, convert kilograms to grams (and vice versa) and add masses given in different units.

 Explain and justify thinking

Classroom technique: One between two

Two children share one pencil. One child must decide on the calculation and explain to their partner why they have chosen this one. They then talk their partner, who has the pencil, through this calculation, saying what to write in order to carry it out.

 Improve learning and performance: assess learning progress

In the plenary, children answer self-assessment questions with the help of their partner, then write a general statement about their progress in this aspect of maths.

Lesson 23: Using a scale plan

 Work with a scale plan

When solving problems involving scale plans, children need to ensure that every element of the problem is to the same scale. Here, children plan how to furnish a room and, in the process, must draw plans of furniture to the same scale as the room.

 Discuss progress of work

Classroom technique: Tell your partner

In class discussion, an alternative to taking answers from individual children is to ask everyone to turn to their partner and tell them the answer to the question. This technique is used in the present lesson to involve all children in thinking and talking about how they will answer the problems posed.

 Organise work: work on a task with several aspects

Children deal with a complex task involving several stages and have to keep track of where they are and what they need to do next. At the end, they need to check they have completed the task as required.

Lesson 24: Using timetables

 Solve problems involving timetables

Planning a journey by public transport is an important life skill. In this lesson, children plan such a journey, working out the length of the journey from a timetable and other information and building in extra time for waiting, changing, traffic jams, and so on.

 Reach a common understanding with a partner

Classroom technique: Talking partners

Children work informally in pairs, talking through and making sense of the problem together, then collaborating on carrying out the calculations and presenting their results.

Organise work: plan ways to solve a problem

Children talk together about the story of the journey, imagining what is involved to help them plan how to solve it.

Area and perimeter

Classroom technique: Talking partners

Learning objectives

Maths
Calculate the area and perimeter of compound shapes

Speaking and listening
'Share ideas and reach agreement'
Share and discuss ideas and reach consensus

Personal skills
'Work cooperatively with others'
Work with others: work cooperatively with others

Words and phrases
constant, compound, area, perimeter, centimetre, square centimetre, distance, edge

Resources
1 cm squared paper (optional)

Introduction

Display two rectangles with their dimensions marked.

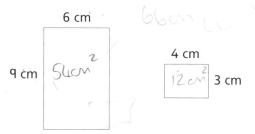

Children find the area and perimeter of each rectangle.
Display a compound shape made up of the two rectangles.

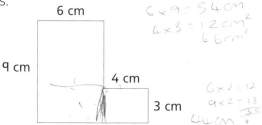

Children now calculate the area and perimeter of the new shape.

Establish a definition with the class for the words 'area' and 'perimeter' and scribe these on the board. Revise the difference between the linear unit 'centimetre' and the unit of area 'square centimetre'.

(m) *Do you think the area of the compound shape is the same as the two separate rectangles? Why would it be the same?*

(S) *Will the perimeter be the same, too? Explain to your partner why/why not.*

Tell your partner
Use this technique (p10). Individual children think about the problems for half a minute, then share their ideas with a partner.

Pairs

Working in pairs, Child A sketches a rectangle (maximum length 12 cm) and marks on it the length and width (using whole centimetres). Child B draws a rectangle beside it, on a common base line, writes in its dimensions and rubs out the section of line separating them.

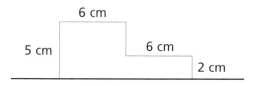

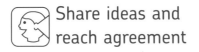
Modelling the way of working
You could demonstrate this on the board with a volunteer to make explaining the task easier.

Children work out the total area as well as the perimeter of the compound shape so made and write this at the bottom of their sheet of paper.

They then make three compound shapes, with the three areas muddled up and, likewise, the three perimeters.

Pairs then swap work with another pair who must match up the shapes with their correct areas and perimeters.

Finally, pairs take back their papers and check the other pair's work.

Finding perimeters
Some children may be able to see that a short cut to finding the perimeter of a compound shape is to imagine it 'filled out' to make one large rectangle.

(m) *If you drew a rectangle with an area of 16 cm and a short perimeter, what sort of shape would it be?*

Support: Children draw their shapes on 1 cm squared paper.

Extend: Children make compound shapes from at least three rectangles.

Plenary

Ask for areas of the compound shapes children created and recreate on the board two which had the same area.

Be prepared
You could have two such shapes prepared in case the children cannot provide them.

Pairs talk about why the perimeters vary, even though the areas are the same. Collect in ideas about this and establish that some of the original perimeter is 'lost' (rubbed out) when the two shapes are combined, although their areas remain the same. Agree a class statement and scribe it.

(m) *If two shapes have the same area, does that mean they have the same perimeter?*

(☺) *What kind of shape might have a small area but a really long perimeter?*

Assessment for learning

Can the children

(m) Find the area and perimeter of simple rectangles? Of compound shapes?

(☺) Turn to their partner and readily discuss the questions you pose in the course of the lesson?

(☻) Take turns to record their work?

If not

(m) Discuss some 'real-life' situations where we need to know the area and/or perimeter of a space and help children devise a mnemonic to remind them of the difference between area and perimeter.

(☺) Give pairs of children practice with this kind of work, using discussion topics that appeal to them. Then tell them you expect similar commitment to discussion when working with other people and other topics.

(☻) Use 'One between two' (p8) to give pairs practice in turn-taking, talking and sharing.

Reading scales

Classroom technique: One between two

Learning objectives

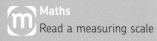
Maths
Read a measuring scale

Speaking and listening
'Explain what you think and why you think it'
Explain and justify thinking

Personal skills
'Assess your progress in learning'
Improve learning and performance: assess learning progress

Words and phrases
mass, weight, weigh, gram, kilogram, decimal, fraction, scale, interval, convert, equivalent

Resources
display copy of RS29
for each pair:
copy of RS30 or RS31

Odd ones out
Some of the amounts are odd ones out, put in to make children think, and do not match with any other.

One between two
Child B has the pencil and writes what their partner asks them to on a separate sheet of paper. Child A thinks out loud, saying why they are choosing the amounts they do. Remind children to take care that they instruct Child B to carry out any necessary conversions, so that the units are the same.

Introduction

Display RS29. Children match pairs or trios of equivalent amounts and give an explanation.

Show pairs of equivalent masses on an adapted place value grid to demonstrate their equivalence. (Convert fractional amounts such as $1\frac{1}{2}$ kg to decimals before putting them on this grid.)

1 kg	$\frac{1}{10}$ kg or 0.1 kg	$\frac{1}{100}$ kg or 0.01 kg	$\frac{1}{1000}$ kg or 0.001 kg	
0 .	7	5		kg
	7	5	0	g
1000 g	100 g	100 g	1 g	

(m) *How do you know that 750 g, and not 7500 g, is a match for 0.75 kg?*

(m) *How many kilograms is 7500 g equivalent to?*

Pairs

Pairs share a copy of RS30, taking turns to find the missing masses marked by letters on the scale.

Child A then chooses a label, reads out the mass marked on it and finds two amounts, one from each list, that total the chosen mass. They instruct Child B how to record the calculation adding these masses.

> A: 500 g
> $\frac{1}{10}$ kg = 100 g
> 100 g + 400 g = 500 g

Children then swap roles and continue working like this until the task is complete.

(m) *How do you work out the masses that the lines are pointing to?*

Explain what you think
and why you think it

(m) *How do you know the value of each interval on the scale?*

(💬) *How can you be sure you have added those two masses correctly?*

Support: Children work with RS31 (a simpler version of the problem).

Extend: Instead of using the masses in the lists, children make up three amounts, in mixed units, that total each mass marked on the scale with a letter.

Plenary

Present these statements for children to discuss with their partner.

> I can put this amount on the scale: 1.8 kg
> I can put this amount on the scale: 450 g
> I can put this amount on the scale: $1\frac{1}{2}$ kg

Assess learning progress

Children think about what they find easy and what they find hard. Writing down a statement in this way helps them take responsibility for their own learning.

Children then write a general statement about their own progress at reading scales which their partner can confirm as accurate.

(💬) *How would you work out where to put $1\frac{1}{2}$ kg on the scale?*

(☺) *Ask your partner which part of today's task they think you found hard.*

Assessment for learning

Can the children

(m) Put on the scale a mass expressed as grams? A mass expressed as vulgar fractions (for example, $1\frac{1}{2}$ kg)? A mass expressed as decimals (for example, 0.25 kg)?

(💬) Explain how they work out the value of a marker on the scale?

(☺) Say what aspects of the work they find easy or difficult?

If not

(m) Do more work with an adapted place value grid to emphasise the relative values of each part of a written mass.

(💬) Ask another child to explain how they do it, then see if the first child can re-explain that child's method.

(☺) Give the child some individual attention, comment yourself on what they seem to be finding easy or hard and ask them to confirm your assessment. It is important that children develop enough self-knowledge to recognise and name how they are feeling about the work they do.

Using a scale plan

Classroom technique: Tell your partner

Learning objectives

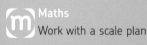 **Maths**
Work with a scale plan

Speaking and listening
'Talk about the progress
of your work'
Discuss progress of work

Personal skills
'Keep track of your work'
Organise work: work on a task
with several aspects

Words and phrases
dimensions, scale,
plan, convert, layout, units
of measurement, equivalent

Resources
display copy of RS32
for each pair:
copy of RS32 or RS33
squared paper
ruler and scissors

Tell your partner
Children turn to their partner and
tell each other what they must
do to answer the questions.
They then work it out together.

Tell your partner
Use this technique again, but tell
children not to start answering
the question.

Mini-plenary
For example, is it helpful to draw
the furniture to scale on squared
paper and cut it out and move it
around before making any
drawing on the final plan?
Does it matter whether the door
opens in or out? Does anyone
need help from a peer tutor on
converting measurements
from millimetres to metres?

Introduction

Display a copy of RS32. Working in pairs, children read
the first two questions, agree how to find the answers,
then find them. (They will need to measure the plan
with their rulers to work them out.)

Establish the correct answers: 4 cm for every 1 m,
or 4:100.

Pairs then share a copy of RS32 and some squared
paper and a ruler.

They look at the third question and discuss what they
will need to do to here.

Collect in suggestions from the children and establish
a procedure: to redraw the plan on squared paper;
to convert the measurements of all the furniture to
the same unit; to find a way to show these items of
furniture on the plan.

Leave some things open – for example, at this stage,
children may or may not want to cut out pieces of
paper to represent the items of furniture.

m *How do you know the scale of the bedroom is
4 to 100?*

What are you being asked to do? And what else?

Why is it a good idea to use squared paper?

Pairs

Pairs work together to plan the bedroom, agreeing
where each piece should be put.

After a few minutes, stop the class and discuss how
the work is progressing.

Five minutes before the plenary, ask children to agree
with their partner how far they have got with the task,
and what they still need to do, and to check the work
they have done so far.

m *Does that shape look right for a table?*

*What will you do if you cannot agree the room
layout with your partner?*

What do you need to do next?

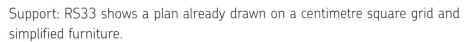
Support: RS33 shows a plan already drawn on a centimetre square grid and simplified furniture.

Extend: Tell children: "You also have a computer table. Decide what its dimensions might be and plan where to put it."

Plenary

If children have not all completed their plan, allow time on another occasion for any further work they need to do.

Meanwhile, help children check they have converted the furniture dimensions correctly, using 'Tell your partner' (p10). Choose an item of furniture and ask children to tell each other its dimensions in metres. Ask an individual for the answer and a brief explanation of how they know this.

(m) *How many metres long and wide is the bed? How do you work that out?*

(m) *If the bed were 10 cm wider, what would the dimensions be in metres?*

Assessment for learning

Can the children	If not
(m) Draw out a plan for a piece of furniture using the same scale as the bedroom plan?	(m) Help the children draw up a conversion table using a constant ratio and use this to draw a piece of furniture to scale:

Room		Plan
1 m	⇒	4 cm
75 cm	⇒	3 cm
50 cm	⇒	2 cm
25 cm	⇒	1 cm

Can the children	If not
Talk about the progress of their work and whether they are experiencing any problems?	Invite other children to talk about their work to provide a model. Tell the child in question to listen to these reports before they tell you two things about how they are getting on with their own work.
Keep track of what they have done and what they still need to do?	Remind children of the procedure that was discussed in the introduction to the lesson and ask them to tell you which of the actions they have taken so far. Then remind them of what they still need to do.

Using timetables
Classroom technique: Talking partners

Learning objectives

m Maths
Solve problems involving timetables

Speaking and listening
'Reach an understanding with your partner'
Reach a common understanding with a partner

Personal skills
'Plan ways to solve a problem'
Organise work: plan ways to solve a problem

W Words and phrases
digital, 24-hour clock, am, pm, arrive, depart, faster, slower, takes longer/less time

r Resources
display copy of RS34
for each pair:
copy of RS34

Ticket to explain
When children need to do several calculations and explain to the class, you can use this technique (p11). A child who wants to offer an explanation must first re-explain the one given by the previous child.

Relevance of the problem
To make the problem more 'real', consider using a destination and train/bus timetables relevant to the children.

Sharing the work
Remind children to share not only the discussion but the doing of the calculations. Children should be prepared to hand over the pencil to their partner at any point in the calculation, and their partner should be able to complete it seamlessly.

Introduction

Display these train times.

| Departs London | 0855 | 1130 | 1215 | 1505 |
| Arrives Leeds | 1118 | 1356 | 1503 | 1730 |

Establish how times are represented on the timetable (0855 is 8.55 am), then ask children to discuss in pairs which trains are fast and which are slow.

Take feedback on how children work out the individual journey times.

m *How did you calculate the length of the journey to Leeds?*

m *If someone has to be in Leeds in time for lunch, which train do they need to catch from London?*

Pairs

Display a copy of RS34 and read through the information (but not the timetable) with the class. Pairs then talk through Teri's journey as a story outline: what she did first, what she did next, and so on.

Take feedback on this, taking the opportunity to help children focus on details such as walking from the train at King's Cross to the bus stop, waiting for the bus, possible traffic delays, getting lost.

Pairs of children now share a copy of RS34 and discuss how to solve the problem presented there. They solve it, then present their results as a schedule for the journey, which Teri could use to get her from home to her destination in good time.

Do you both agree about that?

Talk with your partner about how much time Teri needs to get from the train to the bus stop.

What do you need to do first? How will that help you?

Is that all the information about times you need? Or is there anything else you may need to guess?

Support: Organise pairs so that weaker children work with a competent partner. Ensure that both understand the problem and its solution.

Extend: Teri has no timetable for the journey home, but based on her outward journey, what time should she leave the museum to be home by 6 pm?

Plenary

Help children display their work informally around the classroom. Children then look at each other's work, comparing results (the timings recommended to Teri) and presentations.

Looking at each other's work
This is an opportunity for children to learn, by example, how to present work neatly, clearly and attractively.

At the end, children report back on any features of each other's work that they liked or considered useful.

m *Did anyone suggest catching a different train than the 0927 from Bedford? What were their reasons for catching that train?*

Is there anything you would do differently next time we work on problems like this?

Assessment for learning

Can the children

m Read the timetable and work out the length of the train journey?

Tell you something they agreed/disagreed about when discussing the journey?

Identify a useful thing to do first when presented with the problem?

If not

m Revise with the children how to tackle the problem in stages – for example, make the departure time up to the nearest hour (taking note of how many minutes they are adding on), then count on in whole and part hours from there.

Observe this pair working and check whether the partnership is truly collaborative. If not, consider arranging other partnerships and, perhaps, make more use of techniques such as 'One between two' (p8).

Consider having a class blitz on problem solving and analysing the processes children use for it.

Self and peer assessment

Lesson 21: Area and perimeter	I think	My partner thinks
(m) I can find the area and perimeter of simple rectangles/compound shapes.		
I listen to what my partner says, even if I don't agree with it.		
I make sure my partner and I take turns to record our work.		

Lesson 22: Reading scales	I think	My partner thinks
(m) I can put this amount on the scale: 1.8 kg		
I can explain how to work out the value of a marker on the scale.		
I can say which aspects of this work I find difficult.		

Name _____

Lesson 23: Using a scale plan	I think	My partner thinks
(m) I can draw a plan for a piece of furniture using the same scale as the bedroom plan.		
I can say how our work is going, and whether I am having any problems.		
I keep track of what I have done and what I still need to do.		

Lesson 24: Using timetables	I think	My partner thinks
(m) I can work out the length of a train journey.		
I discuss with my partner if we disagree about something.		
I can find a useful thing to do first when dealing with a maths problem.		

Self and peer assessment

Shape and space

Learning objectives

	Lessons			
	25	**26**	**27**	**28**
ⓜ Maths objectives				
recognise reflective symmetry in 2D shapes	●			
recognise patterns and relationships		●		
describe and classify 2D shapes			●	
describe and visualise 2D shapes				●
Speaking and listening skills				
reach a common understanding with a partner	●			
use precise language to explain ideas or give information		●		
contribute to whole-class discussion			●	
listen with sustained concentration				●
Personal skills				
improve learning and performance: critically evaluate own work	●			
work with others: overcome difficulties and recover from mistakes		●		
improve learning and performance: develop confidence in own judgements			●	
work with others: show awareness and understanding of others' needs				●

About these lessons

Lesson 25: Reflective symmetry

 Recognise reflective symmetry in 2D shapes

Children are often competent at working with reflections when there is a single axis parallel to the edges of their paper. But they also need to work with more complex reflections – for example, where the axis is at an angle or where there are two axes. This activity gives children practice in reflecting shapes in some of these more complex ways.

 Reach a common understanding with a partner

Classroom technique: Telephone conversation

Children have identical grids, but cannot see each other's work. This means they must describe verbally where to place counters so that their patterns are the same and are correctly symmetrical.

 Improve learning and performance: critically evaluate own work

In the plenary, children talk to a partner about their work during the lesson, how they found it and what they achieved.

Lesson 26: Problem solving with squares

 Recognise patterns and relationships

Investigations are open-ended problems with many solutions, and the secret is to find a way to organise these. Children with little practice at investigating may take a muddled approach, but the more experience they get, the smarter investigators they will become.

 Use precise language to explain ideas or give information

Classroom technique: One between two

The challenge is to draw squares and find ways of cutting them up into smaller squares. One child has the pencil, but must follow the instructions given by the other child. This requires them to use the language of position precisely and accurately.

 Work with others: overcome difficulties and recover from mistakes

Children who are struggling or making errors have the opportunity given them by a mini-plenary to check on what other children are doing and learn from them how to tackle the problem.

Lesson 27: Properties of 2D shapes

 Describe and classify 2D shapes

This activity focuses on some of the vague or confused ideas children can have about 2D shapes. In discussing which statement is true, children air these confusions and have an opportunity to clear them up.

 Contribute to whole-class discussion

Classroom technique: Devil's advocate

Statements – false or ambiguous as well as true – can be better than questions at provoking discussion. In this lesson, children are presented with statements to agree or disagree with. Working as a class, children argue for or against a statement, aiming to reach consensus.

 Improve learning and performance: develop confidence in own judgements

Challenging vague or imprecise ideas and establishing the truth can give children confidence that their freshly sharpened ideas are both true and accurate.

Lesson 28: Describing 2D shapes

 Describe and visualise 2D shapes

In this activity, children describe a shape so their partner who cannot see it can draw one the same. In the process, children describe and name 2D shapes and use the descriptions to visualise the shape required.

 Listen with sustained concentration

Classroom technique: Telephone conversation

Because children cannot see each other's work, they must communicate by speech, listening carefully and with concentration to what is said.

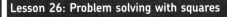 **Work with others: show awareness and understanding of others' needs**

Children have to think about what their partner needs to know, to ensure that the latter draws the shape correctly.

Reflective symmetry
Classroom technique: Telephone conversation

Introduction

Display a copy of RS35 or RS36 and place one counter on the grid. Ask a child to tell you where to put another counter to reflect yours. Repeat until a pattern of several counters is built up.

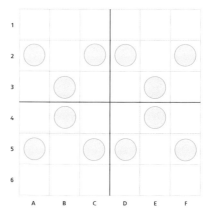

m *How can I check whether the counter is placed correctly?*

Speaking *If you think that counter is in the wrong place, can you tell Megan where you think it should be? And why?*

Counters
Using counters in just one colour means the reflections will be the same colour as the originals.

Pairs

Pairs of children work with a copy of RS36 and some counters. Children sit back to back.

Child A places a counter on one square and another one on its reflections. They tell their partner which squares these are, and the partner covers these squares on their grid.

Leaving the counters in place, Child B has a turn to place counters.

Children have three turns each, then stop and compare sheets.

m *Do those counters look right to you? How can you tell?*

Speaking *How can you ensure that your partner understands what you are saying?*

Reaching consensus
If Child B disagrees about the placing of the 'reflected' counters, the pair discusses this until they are in agreement.

Mirrors
If at all possible, use mirrors to check symmetry. Place a mirror on the line of symmetry and look in it to see what the reflection should look like, then lift the mirror to check whether it does.

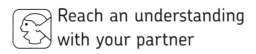
Support: Children work with RS35 and practise first with a grid with a mirror line parallel to the edges of the paper.

Extend: RS37 is challenging as it requires children to find their own way of describing position.

Plenary

Display one or more of the grids and use counters to form some reflections, with the help of the class. Explore and discuss children's methods for checking where a reflected counter belongs.

At the end, children talk about their work on reflection to a new partner.

ⓜ *Tell me where to put the counters to reflect that one.*

ⓜ *How can I be sure those are in the correct places?*

☺ *Think about how clearly you explained to your partner where to place the counters.*

Critical evaluation

Children work with new partners, not the person they did the grids with.

Write up these questions:

– Which grids did you work with?

– What did you do well?

– What did you find difficult?

Children have two minutes each to tell their partner the answers to these questions.

Assessment for learning

Can the children

ⓜ Reflect a counter's position using RS35? RS36? RS37?

☒ Explain to a partner accurately where to place a counter on the various grids?

☺ Identify which grid they are ready to tackle next?

If not

ⓜ Use 'Peer tutoring' (p8) to teach children about working with reflection.

☒ Have children work on these grids with a competent partner, using 'One between two' (p8): the partner models the language, and the added demands of back-to-back work are avoided.

☺ Ask a child's partner to suggest which problems the child could tackle, based on their experience of working with that child, and ask the child if they agree.

Problem solving with squares
Classroom technique: One between two

Learning objectives

(m) Maths
Recognise patterns and relationships

Speaking and listening
'Use precise language when talking'
Use precise language to explain ideas or give information

Personal skills
'Get over difficulties and mistakes'
Work with others: overcome difficulties and recover from mistakes

(W) Words and phrases
investigate, method, check, sort, predict, describe the rule, what could you try next?

(r) Resources
for each pair:
pencil and ruler

Modelling the activity
Work with one or more volunteers to model 'One between two'. As you instruct the child where to draw lines, you can introduce useful vocabulary and ideas about appropriate accuracy.

Squares or not
Children may need to check that their 'squares' really are square. This is an opportunity to focus on properties of squares. For example, if a square is cut into a 3 by 4 grid, can the pieces really be square?

Overcoming difficulties
Mini-plenaries give children who are struggling an opportunity to see how other children are tackling a problem, which can give them a renewed willingness to try.

Counting squares
Children can easily miscount the squares or make inaccurate predictions about the number of squares they are getting, which they fail to check. To avoid this, ask them to number their squares.

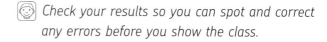

Introduction

Sketch a square on the board and explain that the task is to draw cut lines to divide it into smaller squares. Repeat this with one or two other squares, aiming for different ways of cutting them up into smaller ones.

| 4 squares | 9 squares | 7 squares |

Children count the smaller squares you get each time.

 Draw a line dividing the square in half, from top to bottom.

Do you mean I should divide this small square into four?

Pairs

Pairs of children work together, using 'One between two' (p8), to explore different ways of cutting up squares into smaller ones, aiming to find totals of small squares less than 20. One child has the pencil and ruler, and the other instructs them how to divide up the square.

Stop the class for a mini-plenary. Pairs demonstrate methods they have found to the class. Help them reproduce their sketches on the board.

Pairs then continue the exploration, recording all new successful cuttings.

(m) *Can you make a general statement about what you are finding?*

With all these examples, you have cut a smaller square into four even smaller ones. How many more squares does that add to what you had before?

Check your results so you can spot and correct any errors before you show the class.

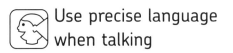
Support: Provide a sheet of ready-drawn squares, each about 10 cm square. Encourage neat and accurate working as well as checking, as these are the best ways of avoiding errors.

Extend: Children look for numbers of squares up to 30. Aim for general statements about what numbers are possible, such as: "Any number of squares you get, you can always get 3 more by cutting one square into quarters."

Plenary

Collect in as many different cuttings as possible. Help the children formulate general statements such as: "If you make a square, you can always cut it into smaller squares."

Work with the class to record systematically all solutions they have found up to 20 or 30.

(m) *What numbers up to 20 can you get? What numbers can't you get, and why?*

(m) *We have cut squares into 2 × 2 and 3 × 3 smaller squares. What would be the next one like that to try? And the next? What kind of numbers are we getting?*

Assessment for learning

Can the children

(m) Make any general statements about which numbers are possible? About which numbers are impossible?

(�) Back up their general statements with reasons?

(☺) Spot errors and correct them?

If not

(m) Help children collect together those cuttings that show square numbers and talk about the numbers these give them.

(�) Ask a more verbal child to provide an explanation and to repeat it.

(☺) Ask children to work with a different partner and check the work together. Ask them to draw out the squares again neatly if inaccurate drawings are hampering the checking process.

Properties of 2D shapes
Classroom technique: Devil's advocate

Learning objectives

(m) Maths
Describe and classify 2D shapes

Speaking and listening
'Join in a discussion with the whole class'
Contribute to whole-class discussion

Personal skills
'Develop confidence about what you think and decide'
Improve learning and performance: develop confidence in own judgements

(W) Words and phrases
isosceles, rhombus, square, trapezium, parallel, oblong, regular, rectangle

(r) Resources
sets of four cards cut from RS38 and RS39
display copies of examples cards from RS38 and RS39
maths dictionary (optional)
wipe boards
pencils

Tell your partner
Children turn to a neighbour and briefly discuss together the properties of the shape in question.

Devil's advocate
In this activity, no one is actually arguing for the false statements (although if children are able and willing to do so, encourage them to). But even being presented with simple statements of contentious ideas provides material for discussion and clarification.

Whole-class discussion
Note down the names of any children who do not offer contributions and ensure they are involved in class discussion on another occasion – perhaps on a more 'comfortable' topic.

Introduction

Brainstorm names of 2D shapes with the class. Write up each shape name suggested and give children a few moments to 'tell their partner' (p10) about the shape.

square	circle
rectangle	equilateral triangle
hexagon	parallelogram
pentagon	isosceles triangle

What is the difference between an equilateral and an isosceles triangle? In what way are they the same?

Tell your partner one fact about squares you are absolutely certain about.

Whole class

Select a set of four cards cut from RS38 or RS39 – for example, set A. Choose three children and give each child one of the statement cards.

They read out their statement twice, but do not say whether it is true or false. The rest of the class decide which statement is correct – only one will be. Give the children a few minutes to discuss with a partner, then ask for a show of hands for each statement in turn.

Finally, show the examples card to the class. Discuss why this proves, or disproves, the other statements and refer to a maths dictionary if appropriate.

(m) *Tell us what you think rectangles are. How can you tell when a shape is a rectangle?*

Does everyone agree with that explanation? Does anyone disagree?

Now you've talked about that statement together, are you confident that it is false?

Does talking together help you get clearer about what something means?

Support: Pair less confident or less articulate children with a partner who has a good grasp of 2D shapes.

Extend: Ask for further statements which might or might not be true and invite the class to discuss them.

Plenary

Look again at the list of shape names from the introduction. Add other words that have come up in the lesson.

Make up incomplete sentences where the missing word is one of those on display. Pairs agree the missing word, write it on their wipe boards and hold it up to you.

Spelling practice
As well as revising the definitions of shape names, you are practising how they are spelled.

Does everyone agree that is the correct word? Is there another word that would fit just as well?

'A straight-sided shape with just one pair of parallel sides is called a ...' Tell us how you know what the missing word is.

Assessment for learning

Can the children

(m) Explain why an isosceles triangle could have a right angle or an obtuse angle?

Offer an idea or comment in class discussion?

Confidently state one or more facts about 2D shapes they are sure about?

If not

(m) Ask children to draw a wide variety of, for example, isosceles triangles – and some triangles that aren't isosceles – and discuss together what their various properties are, to reinforce ideas about shape.

Observe the child in group discussion and note whether they contribute there. If not, arrange for an adult to work with the group and draw out the child so they can experience what it is like to be involved in group discussion.

Go back to basics and offer children simple statements to agree or disagree with. Ask them to practise repeating your statements in a loud, confident voice.

Describing 2D shapes
Classroom technique: Telephone conversation

Learning objectives

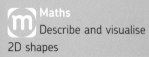

Maths
Describe and visualise 2D shapes

Speaking and listening
'Listen well'
Listen with sustained concentration

Personal skills
'Think about what other people need'
Work with others: show awareness and understanding of others' needs

Words and phrases
vertex, vertices, angle, base, left, right, parallel, regular, triangle, square, rectangle, overlap

Resources
display copy of a squares grid
for each pair:
2 sheets of squared paper
2 rulers
protractor (optional)

Barrier
You could erect a screen between you and the volunteer or simply stand close to your half of the grid so you cannot see their shape.

Describing shapes
Use a version of 'Devil's advocate' (p11): deliberately misunderstand the child's instructions and force them to be more precise about their descriptions.

Introduction

Display a large squares grid and ask a volunteer to draw a simple straight-line shape somewhere on one half the grid – in view of the class but hidden from you. The volunteer describes how to draw the shape so that you can draw one just the same on your half of the grid. They may not see what you draw, but the rest of the class may.

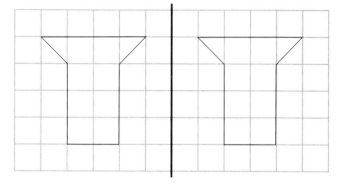

Ask questions as necessary to clarify what is meant. Take the opportunity to introduce vocabulary you want children to use, define it if necessary and write it up.

(m) *Is the triangle sitting flat on its base or is it resting on a point?*

(m) *Class, are our shapes the same? And if not, how do they differ?*

Pairs

Children work in pairs, sitting back to back. Each child has a square grid. Child A draws a shape on their grid, and describes this to Child B, who draws one just the same – same size, same orientation (but same position on the grid may be too hard).

Pairs then swap roles and repeat the activity.

(m) *How can you describe that part of your shape? Can you think of a word that would explain it?*

(💬) *Can you repeat what Ben said?*

(😊) *I wonder if you are giving your partner enough information about the shape.*

RS37

Grid 3

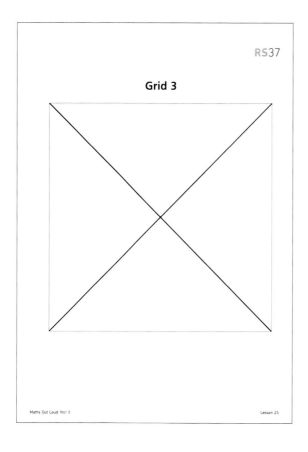

RS38

STATEMENT CARD A1	STATEMENT CARD A2
All isosceles triangles have two sides and two angles the same.	All isosceles triangles have three sides of equal length.

STATEMENT CARD A3	EXAMPLES CARD A
All isosceles triangles have one obtuse and two acute angles.	These are isosceles triangles.

STATEMENT CARD B1	STATEMENT CARD B2
A kite is a special kind of rhombus.	A square is a special kind of rhombus.

STATEMENT CARD B3	EXAMPLES CARD B
A rhombus has three sides of equal length and one side a different length.	These are rhombuses. This isn't.

RS39

STATEMENT CARD C1	STATEMENT CARD C2
A trapezium has all four angles equal.	A trapezium can have one or two pairs of parallel sides.

STATEMENT CARD C3	EXAMPLES CARD C
A trapezium has one pair of parallel sides.	These are trapeziums. This isn't.

STATEMENT CARD D1	STATEMENT CARD D2
An oblong is a regular shape.	All rectangles are regular.

STATEMENT CARD D3	EXAMPLES CARD D
Rectangles are either oblong or square.	Regular shapes have all their sides the same length and all their angles equal.